W9-CBT-933

NOTICE

In the tradition of *Longitude*, a small and engagingly written book on the history and meaning of zero

THE NOTHING THAT IS
A Natural History of Zero
ROBERT KAPLAN

A symbol for what is not there, an emptiness that increases any number it's added to, an inexhaustible and indispensable paradox. As we enter the year 2000, zero is once again making its presence felt. Nothing itself, it makes possible a myriad of calculations. Indeed, without zero mathematics as we know it would not exist. And without mathematics our understanding of the universe would be vastly impoverished. But where did this nothing, this hollow circle, come from? Who created it? And what, exactly, does it mean?

Robert Kaplan's *The Nothing That Is: A Natural History of Zero* begins as a mystery story, taking us back to Sumerian times, and then to Greece and India, piecing together the way the idea of a symbol for nothing evolved. Kaplan shows us just how handicapped our ancestors were in trying to figure large sums without the aid of the zero. (Try multiplying CLXIV by XXIV). Remarkably, even the Greeks, mathematically brilliant as they were, didn't have a zero—or did they? We follow the trail to the East where, a millennium or two ago, Indian mathematicians took another crucial step. By treating zero for the first time like any other number, instead of a unique symbol, they allowed huge new leaps forward in computation, and also in our understanding of how mathematics itself works.

In the Middle Ages, this mathematical knowledge swept across western Europe via Arab traders. At first it was called "dangerous Saracen magic" and considered the Devil's work, but it wasn't long before merchants and bankers saw how handy this magic was, and used it to develop tools like double-entry book-keeping. Zero quickly became an essential part of increasingly sophisticated equations, and with the invention of calculus, one could say it was a linchpin of the scientific revolution. And now even deeper layers of this thing that is nothing are coming to light: our computers speak only in zeros and ones, and modern mathematics shows that zero alone can be made to generate everything.

Robert Kaplan serves up all this history with immense zest and humor; his writing is full of anecdotes and asides, and quotations from Shakespeare to Wallace Stevens extend the book's context far beyond the scope of scientific specialists. For Kaplan, the history of zero is a lens for looking not only into the evolution of mathematics but into the very nature of human thought. He points out how the history of mathematics is a process of recursive abstraction: how once a symbol is created to represent an idea, that symbol itself gives rise to new operations that in turn lead to new ideas. The beauty of mathematics is that even though we invent it, we seem to be discovering something that already exists.

The joy of that discovery shines from Kaplan's pages, as he ranges from Archimedes to Einstein, making fascinating connections between mathematical insights from every age and culture. A *tour de force* of science history, *The Nothing That Is* takes us through the hollow circle that leads to infinity.

Robert Kaplan has taught mathematics to people from six to sixty, most recently at Harvard University. In 1994, with his wife Ellen, he founded The Math Circle, a program, open to the public, for the enjoyment of pure mathematics. He has also taught Philosophy, Greek, German, Sanskrit, and Inspired Guessing. Robert Kaplan lives in Cambridge, MA.

Science/History

$22.00t, 0-19-512842-7, 256 pp., 32 b&w illus., 4¼ x 7¼

8-Copy Counterpack: $176.00t, 0-19-521568-0

THE NOTHING THAT IS

THE NOTHING THAT IS
A Natural History of Zero

ROBERT KAPLAN

New York   Oxford
Oxford University Press
1999

Oxford University Press

Oxford   New York

Athens   Auckland   Bangkok   Bogotá   Bombay
Buenos Aires   Calcutta   Cape Town   Dar es Salaam   Delhi
Florence   Hong Kong   Istanbul   Karachi
Kuala Lumpur   Madras   Madrid   Melbourne
Mexico City   Nairobi   Paris   Singapore
Taipei   Tokyo   Toronto

and associated companies in
Berlin   Ibadan

Copyright © 1999 by Robet Kaplan

Published by Oxford University Press, Inc.
198 Madison Avenue, New York, New York 10016

Oxford is a registered trademark of Oxford University Press

Library of Congress Cataloging-in-Publication Data

ISBN 0-19-512842-7

1 3 5 7 9 8 6 4 2
Printed in the United States of America
on acid-free paper

TO
FRANK BRIMSEK
3 hours 51 minutes 54 seconds

How close to zero is zero?
British Deputy Prime Minister John Prescott

CONTENTS

CONTENTS

ACKNOWLEDGMENTS

First and foremost, the two lighthouses from which I take my bearings: Ellen, whose drawings adorn and whose spirit informs this book; and Barry Mazur, whose verve and insights are endless. This book would have been nothing rather than about nothing had it not been for Dr. Christopher Doyle, Eric Simonoff and Dick Teresi. It has benefitted immensely from the editorial inspiration of Peter Ginna and Stefan McGrath.

There are many in the community of scholars to thank for the generosity of their time and the quality of their knowledge. Jon Tannenhauser has been lavish in his expertise and suggestions, as has Mira Bernstein. Peter Renz, whose store of information is larger even than his private library, has been invaluable. I'm very grateful for their help to Prof. Gary Adelman, Dr. Johannes Bronkhorst, Thomas Burke, Henry Cohn, Professor Paul Dundas, Matthew Emerton, Dr. Harry Falk, Martin Gardner, Nina Goldmakher, Susan Goldstine, Prof. James Gunn, Raqeeb Haque, Prof. Takao Hayashi, Prof. James Rex Knowlson, Michele Jaffe, Prof. Richard Landes, Rhea MacDonald, Dr. Georg Moser, Charles Napier, Lena Nekludova, Prof. Katsumi Nomizu, Prof. Yuri Oda, Larry Pfaff, Donald Rance, Andrew Ranicki, Aamir Rehman, Prof. Abdulhamid Sabra, Dr. Brian A. Sullivan, Daniel Tenney III and Dr. Michio Yano.

Finally, I can't thank enough, for their unfailing support when it most mattered, the Kaplans of Scotland, the Franklins of Wiltshire, the Nuzzos of Chestnut Hill and the Zelevinskys of Sharon.

O

THE LENS

If you look at zero you see nothing; but look through it and you will see the world. For zero brings into focus the great, organic sprawl of mathematics, and mathematics in turn the complex nature of things. From counting to calculating, from estimating the odds to knowing exactly when the tides in our affairs will crest, the shining tools of mathematics let us follow the tacking course everything takes through everything else—and all of their parts swing on the smallest of bvpivots, zero.

With these mental devices we make visible the hidden laws controlling the objects around us in their cycles and swerves. Even the mind itself is mirrored in mathematics, its endless reflections now confusing, now clarifying insight.

Zero's path through time and thought has been as full of intrigue, disguise and mistaken identity as were the careers of the travellers who first brought it to the West. In this book you will see it appear in Sumerian days almost as an afterthought, then in the coming centuries casually alter and almost as casually disappear, to rise again transformed. Its power will seem divine to some, diabolic to others. It will just tease and flit away from the Greeks, live at careless ease in India, suffer our Western crises of identity and emerge this side of Newton with all the subtlety and complexity of our times.

My approach to these adventures will in part be that of a naturalist, collecting the wonderful variety of forms zero takes on—not only as a number but as a metaphor of despair or delight; as a nothing that is an actual something; as the progenitor of us all and as the riddle of riddles. But we, who are more than magpies, feather our nests with bits of time. I will therefore join the naturalist to the historian at the outset, to relish the stories of those who juggled with gigantic numbers as if they were tennis balls; of people who saw their lives hanging on the thread of a calculation; of events sweeping inexorably from East to West and bearing zero along with them—and the way those events were deflected by powerful personalities, such as a little merchant in Pisa; or eccentrics like the Scotsman who fancied himself a warlock.

As we follow the meanderings of zero's symbols and meanings we'll see along with it the making and doing of mathematics—by humans, for humans. No god gave it to us. Its muse speaks only to those who ardently pursue her. And what is that pursuit? A mixture of tinkering and inspiration; an idea that someone strikes on, which then might lie dormant for centuries, only to sprout all at once, here and there, in changed climates of thought; an on-going conversation between guessing and justifying, between imagination and logic.

Why should zero, that O without a figure, as Shakespeare called it, play so crucial a role in shaping the gigantic fabric of expressions which is mathematics? Why do most mathematicians give it pride of place in any list of the most important numbers? How could anyone have claimed that since $O \times O = O$, therefore numbers are real? We will see the answers develop as zero evolves.

And as we watch this maturing of zero and mathematics together, deeper motions in our minds will come into focus. Our curious need, for example, to give names to what we create—and then to wonder whether creatures exist apart from their names. Our equally compelling, opposite need to distance ourselves ever further from individuals and instances, lunging always toward generalities and abbreviating the singularity of things to an Escher array, an orchard seen from the air rather than this gnarled tree and that.

Below these currents of thought we will glimpse in successive chapters the yet deeper, slower swells that bear us now toward looking at the world, now toward looking beyond it. The disquieting question of whether zero is out there or a fiction will call up the perennial puzzle of whether we invent or discover the way of things, hence the yet deeper issue of where we are in the hierarchy. Are we creatures or creators, less than—or only a little less than—the angels in our power to appraise?

Mathematics is an activity about activity. It hasn't ended—has hardly in fact begun, although the polish of its works might give them the look of monuments, and a history of zero mark it as complete. But zero stands not for the closing of a ring: it is rather a gateway. One of the most visionary mathematicians of our time, Alexander Grothendieck, whose results have changed our very way of looking at mathematics, worked for years on his magnum opus, revising, extending—and with it the preface and overview, his Chapter Zero. But neither now will ever be finished. Always beckoning, approached but never achieved: perhaps this comes closest to the nature of zero.

A NOTE TO THE READER
If you have had high-school algebra and geometry, nothing in what lies ahead should trouble you, even if it looks a bit unfamiliar at first. You will find the bibliography and notes to the text on the Web, at www...........

THE NOTHING THAT IS

I

MIND PUT ITS
STAMP ON MATTER

Zero began its career as two wedges pressed into a wet lump of clay, in the days when a superb piece of mental engineering gave us the art of counting. For we count, after all, by giving different number-names and symbols to different sized heaps of things: one, two, three... If you insist on a wholly new name and symbol for every new size, you'll eventually wear out your ingenuity and your memory as well. Just try making up distinct symbols for the first twenty numbers - something like this:

1, 2, 3, 4, 5, 6, 7, 8, 9, Y, /, §, ◊, ℘, ∇, ¬, Ψ, ∧, Υ, λ

and ask: how much is 7 plus 8? Let's see, it is ∇.

And Ψ minus / ? Well, counting back / places from Y, it is 6.

Or γ plus ∧ ? Unfortunately we haven't dreamed up a symbol for that yet - and were we to do so, we would first have to devise seven others.

The solution to this problem must have come up very early in every culture, as it does in a child's life: group the objects you want to count in heaps all of the same manageable, named, size, and then count those heaps. For example,

₥₥ ₥₥ ₥₥ is | | | of the ₥₥ ,

and the unattractive γ + ∧ becomes

|||| |||| + |||| |||| |||| |||,

so |||| of the |||| and ||| more.

The basic heaps tend to have 5 or 10 strokes in them, because of our fingers, but any number your eye can take in at a glance will do (we count eggs and inches by the dozen).

No sooner do we have this short-cut (which brings with it the leap in sophistication from addition to multiplication), than the need for another follows: if |||| |||| |||| + |||| |||| ||| is altogether |||| of the |||| and ||| more, exactly what number is that? Won't we have to invent a new symbol after all? Different cultures came up with different answers. Perhaps from scoring across a stroke like these on a tally-stick, perhaps from hand-signals wagged across the market-place, the Romans let X stand for a heap of |||| |||| , V for |||| ('V', that is, as half - the upper half - of 'X' - a one-hand sign) and so XV for three 5s, on the analogy of writing words from left to right. Instead of the cumbersome VVVV or XVV for four 5s, they wrote XX: two 10s.

So our problem turned into:

X + XVIII = XXVIII.

This looks like a promising idea, but runs into difficulties when you grow tired of writing long strings of Xs for large numbers. At the very least, you're back to having to make up one new symbol after another. The Romans used L for 50, so LX was 10 past 50, or 60; and XL was 10 before 50, so 40. C was 100, D 500, M 1000 and eventually - as debts and dowries mounted - a three-quarter frame around an old symbol increased its value by 100,000. So Livia left 50,000,000 sesterces to Galba, but her son, the Emperor Tiberius - no friend of anyone, certainly not of Galba (and anyway his mother's residual heir) - had D̄ read as plain D, 500 sesterces, *quia notata non praescripta erat summa,* "because the sum was in notation, not written in full." The kind of talk we expect to hear from emperors.

But this way of counting raised problems every day, and not just in the offices of lawyers.

What is 43 + 24? For the Romans, the question was: what is

XLIII + XXIV,

and no attempt to line the two up will ever automatically produce the answer LXVII. Representing large numbers was awkward (even with late Roman abbreviations, 1999 is MCMXCIX:

M	CM	XC	IX
↑	↑	↑	↑
1000	100 before 1000, so 900	10 before 100, so 90	1 before 10, so 9)

but working with any of them is frightening (picture trying to subtract, multiply or, gods forbid, divide).

It needed one of those strokes of genius which we now take for granted to come up with a way of representing numbers that would let you calculate gracefully with them; and the puzzling zero - which stood for no number at all - was the brilliant finishing touch to this invention.

The story begins some 5000 years ago with the Sumerians, those lively people who settled in Mesopotamia (part of what is now Iraq). When you read, on one of their clay tablets, this exchange between father and son: "Where did you go?" "Nowhere." "Then why are you late?", you realize that 5000 years are like an evening gone.

The Sumerians counted by 10s but also by 60s. This may seem bizarre until you recall that we do too, using 60 for minutes in an hour (and 6 x 60 = 360 for degrees in a circle). Worse, we also count by 12 when it comes to months in a year, 7 for days in a week, 24 for hours in a day and 16 for ounces in a pound or a pint. Up until 1971 the British counted their pennies in heaps of 12 to a shilling but heaps of 20 shillings to a pound.

Tug on each of these different systems and you'll unravel a history of customs and compromises, showing what you thought was quirky to be the most natural thing in the world. In the case of the Sumerians, a 60-base (sexigesimal) system most likely sprang from their dealings with another culture whose system of weights - and hence of monetary value - differed from their own. Suppose the Sumerians had a unit of weight - call it 1 - and so larger weights of 2, 3 and so on, up to and then by 10s; but also fractional weights of 1/4, 1/3, 1/2 and 2/3.

Now if they began to trade with a neighboring people who had the same ratios, but a basic unit 60 times as large as their own, you can imagine the difficulties a merchant would have had in figuring out how much of his coinage was equal, say, to 7 2/3 units of his trading-partner's (even when the trade was by barter, strict government records were kept of equivalent values). But this problem all at once disappears if you decide to rethink your unit as 60. Since 7 2/3 x 60 = 460, we're talking about 460 old Sumerian units. And besides, 1/4, 1/3, 1/2 and 2/3 of 60 are all whole numbers - easy to deal with. We will probably never know the little ins and outs of this momentous decision (the cups of beer and back-room bargaining it took to round the proportion of the basic units up or down to 60), but we do know that in the Sumerian system 60 shekels made a mina, and 60 minae a talent.

So far it doesn't sound as if we have made much progress toward calculating with numbers. If anything, the Sumerians seem to have institutionalized a confusion between a decimal and a sexigesimal system. But let's watch how this confusion plays out. As we do, we can't but sense minds like our own speaking across the millennia.

The Sumerians wrote by marking circles and semi-circles with the tip of a hollow reed onto wet clay tablets, which were then preserved by baking. (Masses of these still survive from those awesomely remote days - documents written on computer punch-cards in the 1960s largely do not.) In time the reed gave way to a three-sided stylus, with which you could incise wedge-shaped (cuneiform) marks like this: ; or turning and angling it differently, a 'hook':

. Although the Sumerians yielded to the Akkadians around 2500 B. C., their combination of decimal and sexigesimal counting remained intact, and by 2000 B. C. (Old Babylonian times now) the numbers were written like this:

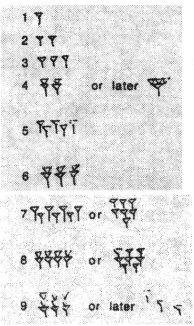

(the diagonal standing for 3 x 3)

7

For 10 they used one 'hook': ⌞

so that 11 was ⌞ ⟑

and 12: ⌞ ⟑⟑

and so on - just like the later Roman invention of X, XI, XII (but there was no new symbol for 5, so 15 wasn't like the Roman XV, but ⌞ ⟑⟑⟑).

20 was ⟨ ⌞ , 30 was made with 3 hooks, variously arranged: ⟨⟨⟨ ⌞

then 40 with 4, 50 with 5; and all the numbers in between were written just as you would expect:

34 was ⟨⟨⟨ ⟑⟑ and 59 was ⌞ ⟑⟑⟑

Now the sexigesimal system intruded. 60 was one wedge again, but a bigger one: ⟑ . So writing numbers, smaller to larger, from right to left (just as we do - thanks to them - although we write our words from left to right), 63 would be ⟑ ⟑⟑⟑ and 72: ⟑⌞ ⟑⟑

You can construct the rest: 120 would be ⟑⟑ , 137: ⟑⟑ ⌞ ⟑⟑⟑

↑ ↑ ↑
(2 × 60) + 10 + 7
= 120

etc. If you want to travel briefly back in time (wet clay, a wooden stylus and a pervasive smell of sheep might help), try writing 217.

Did you get:

⟑⟑⟑ ⟨⟨⟨ ⟑⟑⟑
↑ ↑ ↑
(3 × 60) + 30 + 7
= 180

Notice how important the *size* of the wedges is: the only difference between 62: ⟑⟑⟑ and 3: ⟑⟑⟑ is the larger first wedge. But handwriting - even in cuneiform - changes; people are rushed or careless (try writing up the month's accounts with your stylus), and with thousands of records being kept by harried temple scribes of the names of donors and the number of sheep or fish or chickens each has brought as an offering, the large wedges may grow smaller and the small larger (perhaps from time to time there was a Tiberius factor too), and then where are we?

Utter confusion - until someone comes up with the brilliant idea (or was it a makeshift or compromise that just worked its way into practice, as these things do?) of making the *place* where the wedges are written stand for their value. So large or small, 𒐕𒐕𒐖𒌋𒌋𒐖𒐖 means 202: 3 60s, 2 tens and then 2 more. And 𒐕𒐕𒐕 𒐖𒐖 means 182: 3 60s and then 2.

Once this *positional* system for writing numbers became common practice, it was inevitable that spacing would be called in for clarity, along with stylised groupings of wedges and hooks. So just as our '754' stands for $7 \times 10^2 + 5 \times 10 + 4 \times 1$,

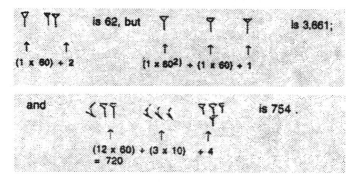

This is wonderful. It not only allows us to write large numbers swiftly

but - much more important - lets us calculate with relative ease. We, for example, add

$$\begin{array}{r} 43 \\ + 14 \\ \hline 57 \end{array}$$

by adding the 3 and 4 first, then the 4 and 1 tens.

For the Babylonians,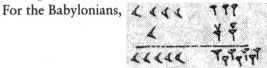

And what about "carrying", one of the sorrows of early childhood?

For us,

$$\begin{array}{r} 82 \\ + 41 \\ \hline 123 \end{array}$$

(2 and 1 units make 3 units, 8 and 4 tens make 12 tens, that is, 2 tens, then 1 hundred). For them,

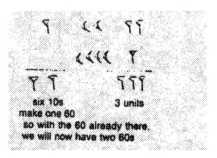

six 10s make one 60 so with the 60 already there, we will now have two 60s   3 units

For us, then, when you move a digit from a column to the one on its left, its value becomes ten times as large - for the Babylonians, 60 times larger. And when one column is full, you empty out its 10 - or 60 - units and put one more unit in the column to the left of it.

No great thing, said Sophocles, comes without a curse. For all the brilliance of positional notation, how were the Babylonians to distinguish between 180: 𒐗 and 3: 𒐗 ? That is, how were they to know whether this '3' was in the units or the 60s column? How are the priests of the temple at E-Mach to know from the records whether two or 120 sheep were given last year as an offering to the goddess Ninmah? Clearly, by context; just as you know where to put the decimal point when you remember that a half gallon of milk costs one fifty five, and your travel agent calls with a bargain flight to Toronto for one fifty five.

But life grows more complicated, the number of things inevitably larger, and context alone becomes a mumbling judge. After putting up with these ambiguities for a thousand years (is it the different rates of change that most tellingly separate cultures?), someone at last - between the sixth and third centuries B. C. - made use of the sign ◁ that had acted as a period, or separated words from their definitions (or in bilingual texts, the transition from one language to another) to wedge columns apart, standing in effect for: "nothing in this column". So

$$𒐖 \quad 𒐙 \quad = 125$$
$$\uparrow \qquad \uparrow$$
$$(2 \times 60) + 5$$

$$\text{but} \quad 𒐖 \quad ◁ \quad 𒐙 \quad = 7{,}205$$
$$\uparrow \qquad \uparrow \qquad \uparrow$$
$$(2 \times 60^2) + (0 \times 60) + 5$$
$$= 7200$$

As you might expect, people had various ways of writing this zero, differing hands differently disposing of what the mind proposed: so

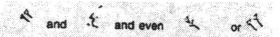

In a tablet unearthed at Kish (dating from perhaps as far back as 700 B. C.), the scribe Bêl-bân-aplu wrote his zeroes with three hooks, rather than two slanted wedges, as if they were thirties; and another scribe at about the same time made his with only one, so that they are indistinguishable from his tens. Carelessness? Or do these errors tell us that we are very near the earliest uses of the separation sign as zero, its meaning and form having yet to settle in?

This zero-marker was used, however, only in the middle of a number, never at its end. It would take a different time, place and people before you could tell from your inventories whether the loaves you had in store would feed 7 or 420.

Still, as circus folk say, what you lose on the roundabouts you gain on the swings. Since you couldn't tell 2 from 20 or 200 without the final zeros, multiplying 2 x 3 or 2 x 30 or 20 x 300 was equally easy: the answer was always 6, with a magnitude that common sense and context made clear. Some even claim that such flexibility was the greatest advantage of this notation.

Whoever it was, in the latter days of Babylon, that first gave to airy nothing a local habitation and a name, has left none himself. Perhaps that double wedge fittingly commemorates his place in history.

2

THE GREEKS HAD
NO WORD FOR IT

Why had it taken so long to signify nothing? Why was the use of zero after that still so hesitant? And why, having surfaced, did it submerge again? The reasons reach down to the ways we turn thoughts and words into each other, and the bemusement this can cause, then as now. Amusement, too: think of the ready pleasure we take in Gershwin's

> I got plenty o' nuttin',
> An' nuttin's plenty fo' me.

We turn over this seeming nonsense with a kind of reflective zest, savoring the difference between what it says and what it means.

A paradox fully as pleasing swept the ancient world. The singers who put *The Odyssey* together, some time near the end of the eighth century B. C., worked into it the story of Odysseus blinding the one-eyed giant Polyphemos, the Cyclops who ate several of Odysseus' crew-members for dinner, and would have devoured the rest had the hero not tricked him.

He got Polyphemos drunk on unmixed wine, and when the Cyclops cried out:

"Give me yet more, and tell me your name right now,
so that I can make you happy by giving you a stranger's
gift,"

Odysseus filled his bowl again and again, and then said:

"Cyclops, you ask for my illustrious name, but I ask that
you give me the stranger's gift just as you promised.
Indeed, my name is Nobody [Οὐτις, Outis]. My mother
and father call me
Nobody, and so do all the rest who are my companions."
Thus I spoke, and right away he answered with ruthless
heart:
"Nobody I shall eat last, after I eat his companions,
and that will be my stranger's gift to you."

Once the Cyclops fell into a drunken stupor, Odysseus
and his men blinded him with a sharp stake, and the giant
gave a horrible cry which brought the other Cyclops all
running. They called to him in his blockaded cave:

"Why are you so overcome, Polyphemos, that you cry
out
through the divine night, keeping us sleepless? It could-
n't be
that some human is leading away your flocks in spite of
you?
it couldn't be that someone is killing you by treachery or
violence?"
From the cave mighty Polyphemos spoke to them:
"O friends, Nobody is killing me by treachery or violence."

His friends, hearing this, went back to their own caves, advising him to bear what the gods send in patience. And so Odysseus and his men escaped, taunting the blinded Cyclops as they rowed away.

You would think that people who could make up and relish such a joke would have no problem with giving a name to nothing, and using the name as cleverly with numbers as Odysseus did with the savage Cyclops. Yet there isn't a trace of a zero in Homeric or Classical Greece - not, in fact, until Alexandrian times, when that glory was past. And if you don't see columns in front of you or in your mind, filling up with counters that then spill over into a single counter in the next column, to leave a blank behind - if you haven't the symbols to stand for those empty or occupied slots, making a *language* out of your deft manipulations - then you won't be able to rise above your handwork, doing what mind does best: taking in and simplifying all that the eye can see - and then moving beyond.

The Greeks of Homer's day just grouped by tens (and at times by fives), using the first letters of the words for these groupings as the number-symbols, and writing down clusters of such signs, from right to left, as the Romans were later to do. So 318, that is,

300 + 10 + 5 + 3, would be

HHH	D	P	III
↑	↑	↑	↑
3xH	+ 1xΔ +	1 x Π +	3

where H, D and P are the first letters of Hekaton (100), Deka (10) and Pente (5) respectively.

No positional notation - hence all of the ills the Romans were later to have with reckoning. Worse: these early Greeks hadn't fully abstracted numbers from what they counted, so that occasionally signs for a monetary unit were combined with those for the amount: instead of HT for one hundred talents (T), they would write |T|. It is as if we were to write $ to mean a dollar and ₿ for eleven dollars, letting the pleasures of doodling lead us to writing as decoration rather than to the peculiarly abstract sort of representation it inclines toward: the making of signs to look through rather than at..

At the height of Greek civilization, in fifth century Athens, a reform swept in for reasons we haven't yet been able to retrieve, which made the 24 letters of the Greek alphabet, supplemented with three more, stand for the first nine digits (1 - 9), then the first nine tens (10 - 90), then the first nine hundreds (100 - 900). So the sign for 10 was ι, iota, the tenth letter of this expanded alphabet, and 11 was ια - but the eleventh letter, κ (kappa) stood for 20. The decimal groupings were now thoroughly disguised. 318, for example, became

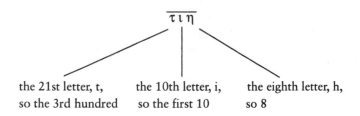

the 21st letter, t,   the 10th letter, i,   the eighth letter, h,
so the 3rd hundred   so the first 10   so 8

The bar drawn over them all was to distinguish this number from a possible word (in this case, τιη , 'why'?). Could the confusion grow greater? It could, and did: at different times, in different places, the order wasn't descending, as here, but ascending; and sometimes all order was ignored. The decorative impulse at play again? Or a code to exclude the uninitiated? Differences that grow from having your neighbors a mountain-valley away? Or just a Greek kind of cussedness of spirit?

Whatever the reason, the continuing lack of positional notation meant that they still had no symbol for zero. It was probably the Greeks under Alexander who discovered the crucial role that zero played in counting, when they invaded what was left of the Babylonian Empire in 331 B. C., and carried zero off with them, along with women and gold. For we find in their astronomical papyri of the third century B. C. the symbol '0' for zero. Why this hollow circle? Where had it come from? The two Babylonian wedges, we know, had a prior, literary existence as separators. We would certainly expect the Greeks to remint this import in their own coinage. Where and exactly when this happened are in all likelihood beyond our reach, the evidence having been pulverized by time. But being human, we can't resist trying to answer the harder and more interesting 'why'. "What song the syrens sang," as Sir Thomas Browne once remarked, "although puzzling, is not beyond *all* conjecture." There is certainly no shortage of conjectures in books as fragile as last year's autumn leaves, and articles buried in morocco-bound mausoleums, and painstakingly-typed manuscript pages in German.

The commonest explanation is that '0' came from the Greek omicron, the first letter of ουδεν, ouden: "nothing", like Odysseus' name Ουτις; or simply from ουκ, 'not': like our nought. The Homeric system, as you saw, drew many of its symbols from the first letters of number-names, and I suppose there is some sort of remote support for this explanation in the fact that ουδεν became μηδεν in later Greek, and a sign a little like μ, Ϥ , is found for 0 in 15th century Byzantine Greek texts.

This whole explanation is curtly dismissed by Otto Neugebauer, the leading authority on Greek astronomical texts, on the grounds that the Greeks had already assigned the numerical value 70 to omicron. The symbol here, he says, was an arbitrary abstraction. Perhaps; but the circle appears at least twice more in Greek mathematics for acronymic reasons. The great Alexandrian mathematician, Diophantus, who flourished in the third century A. D., needed a symbol to separate his ten thousands from lesser numbers, and chose M̊, since 'mo' were the first two letters of 'monad', the Greek for unit. (No one seemed bothered by M̊ also meaning 70,000 - the o for 70 over the M for 'myriad', 10,000 - but then, we get on perfectly well with the same symbol for quotation marks and inches.) And astronomers at the time of Archimedes, four centuries earlier, used μ̊ for 'degree', which in Greek is moira. I find it delightful that the little ° has tripped it down these twenty-two hundred years to remain our own sign for degree.

If you favor the explanation that the '0' was devised by the Greeks without reference to their alphabet, its arbitrariness is lessened by noticing how often nature supplies us with circular hollows: from an open mouth to the faintly outlined dark of the moon; from craters to wounds. "Skulls and seeds and all good things are round", wrote Nabokov.

However the sign for zero evolved, there was always some sort of fancy bar over it: ō̄ or ŏ̄ or ō̄, and even, at times, ⌐o⌐ . These decorations allowed an astronomer like Ptolemy, for example, around 150 A. D., to keep his notation straight. For we find in his *Almagest* ("The Greatest Synthesis"), a "0" both in the middle and at the ends of the three-part numbers he used for his astonishing forays into trigonometry (with degrees, minutes and seconds which he calculated in the Babylonian sexigesimal system, as we do).

So μα ο̄ ιη̄   stood for 41° 00' 18", and σ̄ λγ̄ δ̄

for 0° 33' 04". Doesn't the ornamented bar show that zero hadn't yet the status of a number but was used by the Alexandrian Greeks as we use punctuation marks?

Further evidence of this comes from the sort of complication that keeps scholarship alive. For the only manuscripts we have of the *Almagest* are Byzantine, long post-dating Ptolemy and conceivably influenced, therefore, by intervening practice. And in these Byzantine copies the bars over the letter-names for numbers remain, but that over the zero often disappears - so that zero was still (but now differently) distinguished from number symbols.

Even more to the point is that this 0 indicates the absence of a *kind* of measure (degrees, or minutes, or seconds), but can't yet be taken with other numbers to form a *number*. If you have 38 eggs you would probably say that you had three dozen eggs and two left over - but you'd hardly say you had two dozen eggs and 14 more, even though this would be mathematically correct. So too with English money, before decimalisation, it would have been idiomatically wrong to say you had five pounds, 22 shillings and 14 pence: what you "really" had was six pounds three shillings tuppence. Both of these ways of measuring are throwbacks to our much earlier way of dealing with numbers, where they still acted as adjectives modifying different kinds of heaps. You reckoned the number of elements in a heap only up to certain conventional limits. There was still a long way to go from the key insertion in writing of a sign for "nothing in this column" to such symbols as '106' or '41.005°' (the "numerical" form of 41° 00' 18").

Why didn't the Greeks pursue this way - for the zero hardly appears outside their astronomical writings? And why, after all, hadn't such an inventive people come up long since with writing numbers positionally, and the zero this entails? Why, at the peak of their power, did they step, as you have seen, yet further away from what would have aided thought?

You might argue that it all went back to their admiration of the Egyptians, who had neither a zero nor a positional way of writing numbers. For the Greeks, however, admiration always turned into rivalry; their impatience with the inelegant (and the Egyptian number system lacked any elegance) led to that endless messing about we enshrine as The Scientific Spirit; and curiosity begot ingenuity from dawn to dusk. It still does: I once casually asked a Greek friend in Paris how many of his countrymen lived there. He shrugged. "Who knows? But I'll quickly find out." He leapt from our café table and ran to the nearest wall, where he began to drill with his finger. "What are you doing?" I asked, utterly perplexed. "I don't know," he said, "but Greeks are curious, and soon every Greek in Paris will be here, asking questions and giving advice."

Then if it wasn't homage paid to the stasis that was Egypt, what accounts for this odd kink in Greek intellectual history? Strangely enough, counting hadn't much prestige among them. It was something they called logistic, and tradesmen did it. Not that all the Greeks scorned commerce: they were very good at it, as the extent of the Athenian empire testifies - but their leisure class did, and the thinkers whose writings we have were of this class. Their mathematical energy went largely into geometry, with profound results.

These geometers - in whose midst eventually Socrates and Plato arose - did their arithmetic geometrically, with figures drawn in the sand. 1, 3, 6, 10 and so on were triangular numbers, because you could form them by enlarging equilateral triangles with new rows of dots:

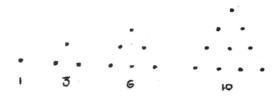

1   3   6   10

This kind of figuring led to deep insights - to sequences of 'square' and 'pentagonal' numbers, for example -

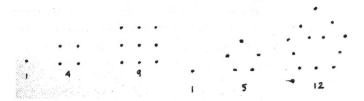

but not to zero or a need for zero, since they weren't calculating but hoping to see how the laws of the universe would be revealed in these combinatorial shadows cast by the play of forms.

Where did this leave the merchants? With a device that the philosophers never described, but whose descendants you see to this day in the worry-beads of the Greeks and the backgammon games of their taverna: the counting board. And even before this - though these boards go back at least to the 7th century B. C. (the Babylonians may even have used them a thousand years earlier) - they had their fingers, and clever ways of flying through calculations with them (you will still see women bending and braiding their fingers in the market, using what some of them call "women's arithmetic", others "the arithmetic of Marseilles").

Solon, the great law-giver of ancient Athens, compares in one of his poems a tyrant's favorite to a counter whose value depends on the whim of the person pushing it from column to column. This metaphor tells us something crucial about zero - especially when we see the metaphor enlarged five centuries later by the historian Polybius:

The courtiers who surround kings are exactly like counters on the lines of a counting board. For, depending on the will of the reckoner, they may be valued either at no more than a mere eighth of an obol, or else at a whole talent [i.e., about 300,000 times as great].

Not "valued at nothing", you notice: for of course there was no column on a counting board that stood for zero (some Greek boards apparently had columns from left to right for a talent (6000 drachmas), then 1000, 100, 10 and 1 drachmas, followed by 1, 1/2, 1/4 and 1/8 obols (6 obols to the drachma), since these were boards for reckoning money, like the boards in those medieval counting-houses that turned into our "banks", i.e., counting-benches). In other words, "nothing" wasn't a *thing*, a *number*, but a *condition* - often transitory - of part of the board. So too in finger-reckoning, if we go by a much later system explained by the Venerable Bede in his "On Calculating and Speaking with the Fingers", written about 730 A. D. There 'zero' was indicated by the relaxed or normal position of the fingers: in other words, it was no gesture at all.

The people who did the calculating among the Greeks, then, hardly needed names for their amounts, since they could *show* them by piles of counters in columns:

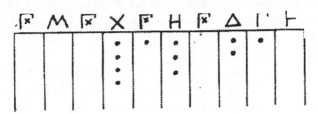

4825: 4 counters in the X (= 1000) column, so 4000
1 counter in the (= 500) column, so   500
3 counters in the H (= 100) column, so   300
2 counters in the Δ (= 10) column, so   20
1 counter in the (= 5) column, so   5
                                             ————
                                             4825

Further manipulations would simply be seen. But even to the extent that they needed names, 'zero' wouldn't be among them.

Digress with me for a page or two: these counting boards open up some lines of conjecture about the source of zero's symbol. On the Darius vase - an elaborate red-figure krater from the fourth century

A treasorer reckoning (from the Darius vase)

B. C. found in Apulia - the Royal Treasurer is seen calculating the value of the tributes paid in by the conquered nations, whose representatives crouch before him. He sits at a reckoning table with signs for monetary values on it, one of which is O, the Boeotian for 'obol', a coin worth, as you saw, almost nothing (they shipped off the dead with an obol under the tongue to pay the eternal ferryman). Almost nothing? If you haven't a symbol at hand for nothing itself, would one for a value close to it serve? It did in a Coptic text where a noon-time shadow in June, near the equator (no shadow at all, that is) is said to be half a foot long, only (according to its modern editor) to avoid having to say 'zero'. Or would Hamlet's friend Horatio tell us that 'twere too curious to consider so? Personally I think that not the column names but the counters on these boards give us the clue we've been looking for. Surely the pebbles used as counters would have been more or less round, hence reasonably represented in writing or drawing by solid dots: •. A graphic way of showing that not even a single counter was in a column would therefore have been an empty dot: o. Not a long step, then, from a drawing to a symbol (a figure to a figure): think of that neat piece of criminal cant, "Giving a place the double-o", for <u>o</u>nce-<u>o</u>ver - but also with a visual pun on a pair of watching eyes. And why did this round o elongate over the centuries to 0? Because split quills and pen nibs make a continuous circle harder to draw than two vertical, curved strokes.

If the step from signalling presence to absence, • to o, still seems too long for you, here is a companion conjecture fetched from not very far to abridge it. Might not the Greek geometers have hit on their sequences of triangular, square and polygonal numbers by playing impishly around with the merchants' counters? They wouldn't have been alone in putting these markers to original uses: two millennia later the young Goethe delighted in rearranging the stones on his

father's counting board into the shapes of the constella-
tions. Now we gather from Plato that the geometers drew
their figures at least some of the time in the sand. If they
built up their shaped numbers with pebbles in the sand too,
then the merchants - or whoever used the counters for reckoning - seeing these patterns would have seen too the result
of taking a counter away: a circular depression left in its
place, o for •. Is that Horatio clearing his throat again at all
these ifs? We could satisfy him in the unlikely event that we
came on counting boards sprinkled with sand.

Climbing back in from this limb, we can agree that there
was one way of writing numbers for the Greeks, and
another way of reckoning with them. The snobbery of the
Athenian aristocrats can't wholly account for this divergence: something deeper, as well as darker, was at work -
that issue of words and thoughts turning into each other.
Watching pebbles being moved quickly around in front of
you is unlikely to inspire trust - as any victim of the old
shell game will tell you. Aristophanes, who wrote his come-
dies in fifth century Athens, had one of his characters in
The Wasps say that the city's finances should be worked
out not by "pebbling" but with fingers. Yet why should finger-reckoning be any more trustworthy? It leaves no per-
manent record. A code was needed both flexible enough to
keep up with and aid the mind, yet sufficiently foolproof to
resist the likes of Tiberius. You notice we have still to solve
this problem fully: on your checks you must write out the
amount in words as well as in figures and the bank will
take the words over the figures, should these disagree, as
being less susceptible to forgery.

A way to calculate yet keep a clear record: this is where
body and mind diverge. Think of the thousand nameless
actions that fill the crevices of your day: modulating your
voice to convey interest or disdain; tying your shoelaces;
whipping up an omelette or flipping an accurate throw.

These are the moves your body knows but would stumble over were you to try describing them. Yet it isn't until these manoeuvres make their way, however shyly, into speech that we can abstract from them and so bring them into the theatre of thought. Zero - balanced on the edge between an action and a thing (and what are numbers, when it comes to that: adjectives or nouns?) - perplexed its users whenever they stopped to think about what they were doing.

There is a last veil that the Greeks may have drawn over this number that doesn't number: it was a veil we know that they drew over much else. Language falls between our acting and our thinking; but language itself has two layers, the spoken and the written. The permanence of writing has always made it the more valuable of the two for us, even at the cost of trading in slang for solemnity. Yet not quite always: the Greeks of that golden age had peculiar views, some of them based on the remarkable ability of their singers to know vast epics like the *Iliad* and *Odyssey* by heart. Memory was often equated with knowledge, knowledge with wisdom - so that the external memory of texts (that repository of our culture, binding us to generations gone) must have been for them something like musical scores: you feel a bit let down when a concert pianist has to perform with one in front of him. Perhaps this was why Plato wrote dialogues: they were and were not to be taken at their word. Certainly he deliberately undermines his enterprise in one of them, for in the *Phaedrus* he has Socrates argue that writing will cause forgetfulness and give only the semblance of truth. This may also be why that earlier philosopher, Heraclitus, made his aphorisms short and perplexing , and why in fact the Greeks invented irony, where you mean only some of what you say but don't say most of what you mean.

And zero? Its singular absence from Greek texts may not show they hadn't used or thought about it: indeed, perhaps the very opposite. Secrecy shrouded the doings of the Pythagorean fraternity living in their midst: mathematics was what mattered for them and its initiates kept to themselves its revelations of cosmic order (within the hierarchy some knew the further mysteries of the disorder threatened by irrational numbers). Could they have been the custodians therefore of some secret traditions involving zero, beginning that long slide out of sight that it was to suffer, emerging only centuries later in the heat - and especially dust - of India?

Of course evidence of the sort that the dog didn't bark can never be admitted into the courts of history. There the standards of proof hold us to reading the lines, not between them. But mind delights too in finding directions out by indirections, and nods are as good as winks outside of chambers: they alert you to whatever signs might come from this quarter, and after all suit the absence that zero stands for.

3

TRAVELLERS' TALES

What happened in that autumnal world, long ago, as the thought of Athens shifted to Alexandria, its power to Rome and its culture, carried eastward by invasion and trade, fchanged in new surroundings while those surroundings changed to absorb it? We are past the days when geometry snubbed arithmetic, and so would expect to find zero coming into its own. Here was this symbol with immense power to describe, explain and control locked up in its little ring, being passed from language to language, from one mathematician or astronomer to another, with none realizing what he had in his pocket.

As in all the best adventure stories, it didn't turn up where it should have: in Sicily, for example, when during the third century

B. C. a passion for huge numbers blossomed. You would think that botanising among such growths would lead inevitably to a full positional notation and zero, its *genius loci:* not different kinds of heaps present or absent, but numbers as such, written positionally and abstracted from what they counted. You've seen how hard it is to think up and manipulate new names for ever larger ensembles - but how easy to add another zero to a row stretching out after a harmless "1". This is certainly how we picture the zillions

Archimedes about to be killed by an impatient centurion

invoked to express awe or desire. In my neck of the woods, where we tried to outdo each other as kids with bazillions and kazillions, it always came down to who could squeeze one last zero onto the page - like the barmen of Dublin who always manage to fit yet one more drop of Guinness into a brimming pint (what a history could be written of our reaching toward the infinite, and the fitful evolution of fantasy to imagination, by looking at our changing ways of naming vast multitudes). Yet an inventor who gloried in mind-boggling numbers did so with never a zero in sight.

Archimedes was born around 287 B. C., the son of an astronomer. Among his amazing works is one he sent to Gelon, King of Syracuse, in which he shows how to name quantities greater not only than the number of grains of sand on all the beaches around Syracuse, but on all the beaches of Sicily; and in all the lands of the world, known or unknown; and in the world itself, were it made wholly of sand; and, he says, "I will try to show you by means of geometrical proofs which you will be able to follow, that, of the numbers named by me..., some exceed the number of the grains of sand... in a mass equal in magnitude to the universe."

This monstrous vision, which puts in the shade such fairy-tale excesses as the mill grinding out salt forever on the floor of the sea, Archimedes makes precise by an ingenious sequence of multiplications.

Take it, he says, that there are at most 10,000 grains of sand in a heap the size of a poppy-seed; and that a row of 40 poppy-seeds will be as wide as a finger. To keep things simple, picture each seed as a sphere. Since the volumes of spheres are to each other as the cubes of their diameters, this line of 40 seeds becomes the diameter of a sphere with a volume $(40)^3 = 64,000$ times the volume of one seed; and since that one holds 10,000 grains of sand, we're already talking about 64,000 x 10,000, that is, 640,000,000 grains. In our modern notation, that's $4^3 \times 10^7$ grains. Round 64 up to 100 for convenience, and we'll have 10^9 grains in a sphere whose diameter is a finger-breadth. Don't worry that all these estimates may be too large: exaggeration, as you'll see, is part of Archimedes' game.

Now 10,000 (10^4) finger-breadths make a Greek unit of length called a stade (roughly a tenth of our mile). These spheres, whose diameter is 10^4 finger-breadths, will have a volume $(10^4)^3 = 10^{12}$ times the volume of one, and that one contains 10^9 grains of sand; so a sphere with a one-stade diameter will contain $10^{12} \times 10^9 = 10^{21}$ grains of sand.

Archimedes next draws on the work of a great astronomer some 25 years his senior, Aristarchus of Samos, to estimate the diameter of the universe (which for the Greeks meant out to the sphere of the fixed stars). Aristarchus - so long before Copernicus! - held that the earth circled the sun. Archimedes juggled with Aristarchus' observations and calculations to come up with an imaginary sphere (call it S) whose radius is the distance from earth to the sun; then he assumes that

$$\frac{\text{the diameter of the earth}}{\text{the diameter of S}} = \frac{\text{the diameter of S}}{\text{the diameter of the universe}}$$

This gives him (after modifying Aristarchus' figures)100,000,000,000,000 or 10^{14} stadia for the diameter of the universe. Its volume is therefore $(10^{14})^3 = 10^{42}$ times the volume of the sphere whose diameter is one stade, which held 10^{21} grains of sand. Hence there would be 10^{21} x $10^{42} = 10^{63}$ grains of sand in a universe compacted wholly of sand.

"I suppose, King Gelon," says Archimedes, "that all this will seem incredible to those who haven't studied mathematics, but to a mathematician the proof will be convincing. And it was for this reason that I thought it worth your while to learn it."

When you consider that in the 1940s two persistent New Yorkers estimated that the number of grains of sand on Coney Island came to about 10^{20}; and that present estimates for the total number of much smaller particles in our much larger universe weigh in at between10^{72} and 10^{87}, you have to say that Archimedes' estimate wasn't all that bad.

This is a spectacular application of the Greek insight that the world afar can be grasped by analogy to the world at hand. But it is made much more spectacular when you realize that Archimedes hadn't our convenient notation for powers of ten, all built on the use of zero.

"How could he have missed it?" asked one of the greatest of all mathematicians, Karl Friedrich Gauss, who very much admired Archimedes. "To what heights science would have risen by now," he wrote in the 19th century, "if only he had made that discovery!" But the fact remains that Archimedes worked with number *names* rather than digits, and the largest of the Greek names was "myriad", for 10,000. This let him speak of a myriad myriads (10^8), and he then invented a new term, calling any number up to 10^8 *a number of the first order*.

He next took a myriad myriads as his unit for *numbers of the second order*, which therefore go up to 10^{16} (as we would say - but he didn't); and 10^{16} as the unit for *numbers of the third order* (up to 10^{24}), and so on; so that the unimaginably gigantic 10^{63} is a number somewhere in the eighth order.

But Archimedes doesn't stop there. In fact, he has hardly begun. In a spirit personified for us by Paul Bunyan, he leaves the sand-filled universe behind, diminished itself to a grain of sand, as he piles up order on order, even unto the 10^8 order, which frighteningly enough contains all the numbers from $(100,000,000)^{99999999}$ to $(100,000,000)^{100,000,000}$.

Are we done? Hardly. All of those orders, up to the one just named, make up *the first period*. If you look at his own words your mind begins to go out of focus, and you feel like Alice as she fell toward Wonderland: "Do cats eat bats? Do bats eat cats?" For he says:

> And let the last number of the first period
> be called a unit of numbers of the first order
> of the second period. And again, let a myriad
> myriads of numbers of the first order of the
> second period be called *a unit of numbers of
> the second order of the second period*. Similarly -

Perhaps Gelon had stopped reading by now, and so missed Archimedes' eventual conclusion:

> And let the process continue up to a myriad-
> myriad units of the myriad-myriadth order

of the myriad-myriadth period

(αι μυριακισμυριοστας περιοδου μυριακισμυριοστων αριθμων μυριας μυριαδας)

- or in our notation, up to $10^{80,000,000,000,000,000}$. Of course there wouldn't be enough grains of sand in his universe or ours to trickle this number out, nor enough time, from the Big Bang to now, to recite its digits at one a second, since the last number of his first period is 1 followed by 800 million zeroes, and this one has 10^8 times as many.

What was Archimedes after in all this, and - if it makes sense to shape negative history, speculating about paths not taken - why was zero absent from his contrivings? Some say that his *Sand-Reckoner* was a tour de force, showing that the stylus was mightier than the xiphos. You might think of it as thoroughly Greek in its playfulness: for Plato said we are but playthings of the gods, and so should play the noblest games - and this exuberant work of his, having no conceivable practical use, must be a scherzo. Was it his intention to humble a king, or to glory in surpassing the magnitudes of his predecessors? His father Phidias, for example, had declared that the sun's diameter was 12 times that of the moon. Archimedes took it to be 30 times (he would have been pleased to learn it is actually 400 times as great). And Aristarchus made use of the awesome 71,755,875 in one of his calculations, which Archimedes here strode past in a single step. Was there a kind of chest-thumping rivalry among the mathematicians of his day, as with children of ours, in conjuring up larger and larger numbers? Archimedes' contemporary Apollonius seems to have responded to <u>The</u> <u>Sand-Reckoner</u> with a system of his own for naming large numbers, which Archimedes then countered with a problem whose answer (could Archimedes possibly have known it?) was a number so large that to write it out in digits would take up the next 47 pages (for Archimedes, the answer would have begun: "7 units of 3 myriad 5819th numbers, and 7602 myriad 7140 units of 2 myriad 5818th numbers, and...") You will be amused - or perplexed - to learn that mathematicians still thus see and raise one another, but now with infinities.

More profoundly, wasn't he showing us how to think as concretely as we can about the very large, giving us a way of building up to it in stages rather than letting our thoughts diffuse in the face of immensity, so that we will be able to distinguish even such magnitudes as these from the infinite? As a mathematician I know said recently, "Large numbers are actually very large."

Or are we seeing here that play we have already watched between language and thought, this time not leading to but deliberately avoiding the convenience of zero? For at the beginning of *The Sand Reckoner* Archimedes made, you recall, a curious remark:

> There are some who... think that no number has been named which is great enough to exceed [the number of grains of sand in every region of the earth]. But I will try to show you... that, of the numbers *named* by me... some exceed... the number [of grains of sand that would fill up the universe].

Why does the emphasis fall on naming? Think for a moment of the letter St. Paul wrote to the Ephesians, where he speaks of Christ as

> far above all principality, and power, and might, and dominion, and every name that is named, not only in this world but also in that which is to come.

Haven't we all an ancient sense that for something to exist it must have a name? Many a child refuses to accept the argument that the numbers go on forever (just add one to any candidate for the last) because names run out. For them a googol - 1 with a hundred zeroes after it - is a large and living friend, as is a googolplex (10 to the googol power, in an Archimedean spirit). A seven-year-old of my acquaintance claimed that the last number of all was 23,000. "What about 23,000 and one?" she was asked. After a pause: "Well, I was close." Under this Adam Impulse people have exerted themselves to come up with names for very large numbers, such as primo-vigesimo-centillion for 10^{366}, and the mellifluous milli-millillion for $10^{3,000,003}$. What points to a fundamental trait of ours is that 10^{63}, or 1 with 63 zeroes after it, leaves the imagination cold: it might just as well have had a few dozen zeroes more, or a couple less. What facilitates thought impoverishes imagination.

By not using zero, but naming instead his myriad myriads, orders and periods, Archimedes has given a constructive vitality to this vastness - putting it just that much nearer our reach, if not our grasp. There are other ways, to be sure, of satisfying the Adam Impulse: invoking dread, for example, rather than awe. Eighteen hundred years after Archimedes, John Donne, haunted and haunting, said in a Lenten sermon:

> Men have calculated how many particular
> graines of sand, would fill up all the vast
> space between the Earth and the Firmament:
> and we find, that a few lines of cyphers will
> designe and expresse that number.... But if every
> grain of sand were that number, and multiplied
> again by that number, yet all that, all that inex-pressible,
inconsiderable number, made up

not one minute of this eternity; neither would
this curse [of God on the inveterate sinner] be a
minute shorter for having been endured so many
generations, as there were graines of sand in that
number.... How do men bear it, we know not; what
passes between God and those men, upon whom the
curse of God lieth, in their dark horrours at
midnight, they would not have us know... This is the
Anathema Maranatha, accursed till the
Lord come; and when the Lord cometh, he cometh
not to reverse, nor to alleviate, but to ratifie and
aggravate the curse.

4

EASTWARDS

I have suggested that abstract thought and imagination are rivals; yet why must one prosper at the other's expense? Counting and naming have been twinned as far back as you care to look. There they are in Homer's catalogue of ships:

> ...they who lived in Hyria and in rocky Aulis,
> in Eteonos of the glens, and Schoinos, and Skolos,
> Thespeia and Graia, and in wide Mikalessos...

Even our words for these two acts of mind are parallel: we tell stories as well as beads, tally our accounts and recount our tales.

Mathematics rises from numbering to making relations among quantities, and surely it must benefit from our packing as much as we can into the names it thus relates, filling each with the past and the possible, letting the sounds store up what we know and dream. Then when a new structure is built by combining them, its rooms will be sumptuous, not cells.

The problem is that to focus on relations we must shrink to mere points what they connect - then symbolize those connections in turn to make yet more extended constructions. Enliven the nodes too much and the net will collapse. No lingering among past parts, then, but leaping to the new whole. This is the recursive abstracting which is the very stuff of mathematics, this abbreviating the sweep of landscape you have just taken in to an aperçu for a higher order of seeing. Little wonder, then, that Goethe compared mathematicians to Frenchmen. "Whatever you tell them," he said, "they translate into their own language and all at once it is something completely different."

They are speaking about connections between connections, and the numbers those connections first dwelt among have by then a wraith-like existence. They were fairly tenuous to start with: if you say there are seven apples in a bowl, exactly what does that "seven" belong to? Not to any one of the apples taken singly (not even the last you counted, since you could have arranged them differently), nor to the bowl that contains them, but - to there being just seven of them. Many a fine head has broken on this problem. Some have ended up saying that seven is the set of all those sets that contain seven objects. And if you eat one of the apples, where has the seven gone? Fled, presumably, to those sets that still or newly have seven members.

The situation is even worse with zero. Names belong to things, but zero belongs to nothing. It counts the totality of what isn't there. By this reasoning it must be everywhere with regard to this and that: with regard, for instance, to the number of humming-birds in that bowl with seven - or now six - apples. Then what does zero name? It looks like a smaller version of Gertrude Stein's Oakland, having no there there.

"I can call spirits from the vasty deep", says Owen Glendower in Shakespeare's *Henry IV*. "Why so can I, or so can any man," answers Hotspur, "but will they come when you do call for them?" We can try to call up the spirit of numbers by naming them, but they remain as elusive as ever, with the will o' the wisp of zero dancing them away.

Follow this dance into India along the invasion route of Alexander in 326 B. C., and the later routes of commerce from Alexandria: Greeks bearing the Babylonian gift of zero. We come into a country where gigantic stretches of time and space and frightening numbers of creatures are the commonplaces of myth and folklore. A column of ants, four yards wide, streams across the floor of the palace of Indra, King of the Gods. "What are they?" he asks in awe of the ten-year-old boy who stands as a pilgrim before him. Each was once an Indra, says the boy: Indras who ruled in the countless universes that float like delicate boats side by side in endless space; and each Indra lives 71 eons, and the lives of 28 Indras are a day and a night in the life of Brahma, which is made up of 108 years of such days and nights; and before and after each Brahma is another - Brahma without end.

But when it comes to the pedestrian matter of dating such stories or tracing their antecedents, we must give it up. An attitude more poetic than ours toward when events occurred, and toward the events themselves, makes hazy chronicles of these distant times. Even an early edition of the *Surya Siddhanta* - the first important Indian book on astronomy claimed the work to be some 2,163,500 years older than it has since been shown to be (though this revising wasn't made in time to excuse Christopher Marlowe, accused of atheism partly for pointing out that Indian texts pre-dated Adam).

Can we say that Archimedes' *Sand Reckoner* and its zeroless ranks of numbers influenced a charming story in the *Lalitavistara*, written at least three hundred years later? This life of the Buddha shows him as a young man competing for the hand of Gopa and easily defeating his rivals in wrestling, archery, running, swimming and writing. Then

The Buddha, calculating

comes the examination in mathematics: he must name all the numerical ranks beyond a koti (ten million, i.e., 10^7), each rank being a hundred times greater than the last. Gautama answers: ayuta, niyuta, kankara, vivara, achobya, vivaha, utsanga, bahula, nagabala, titilambha, vyavaithanaprajnapti (! that's 10^{31}), and so through the alluring samaptalambha (10^{37}) and the tongue-twisting visandjnagati (10^{47}) to tallakchana ($10^{7+46} = 10^{53}$) at last.

But it isn't the last after all: just as with Archimedes, this is a first level only. A second takes him up to $10^{7+2\times46} =$

10^{99}, and eventually (while the courtiers in their robes and ornaments stand astounded) the ninth brings him to $10^{7\,+\,9\times46} = 10^{421}$.

For extra credit he names all the atoms in a yojana (a league: roughly three miles): 7 of the finest atoms make a grain of very fine dust, 7 of which make a little grain of dust. 7 such grains make a mote you can see in a sunbeam, 7 of these a rabbit's grain, 7 rabbit's grains a ram's grain, 7 ram's grains an ox's grain, 7 ox's grains - a poppy-seed! Sound familiar? Compiling by sevens he goes on to mustard-seeds and barleycorns and knuckles, twelve of which make a span, two spans a cubit, four cubits a bow, a thousand bows a cry in the land of Magadha, and four such cries a yojana - or $384,000 \times 10^7$ atoms in all (and he goes on to number all the atoms in all the lands of the world, known and unknown, and even in the three thousand great thousands of worlds to which for some reason the universe has shrunk).

Gautama's reward is not only the hand of Gopa but the fulfillment of every schoolboy's dream: the examiner prostrates himself before the youth and exclaims: "You, not I, are the master mathematician!"

Traveller's tales must have a moral, and the better ones have three. A moral of this one is that working your way up to preposterously large numbers serves not only to extend imagination but as a vehicle for reverence - a way of saying: "There were giants in the earth in those days." Reciting this litany of number-names (ayuta, niyuta, kankara, vivara...) puts you in touch with vast, invisible powers, conferring on your telling an incantatory, almost a magical, power. What if some of the names are confused - ayuta and niyuta here are 10^9 and 10^{11}, while elsewhere they are 10^4 and 10^5; what if in other accounts it takes three invisible atoms (or for the sharper-sighted, 30) to make a mote of dust, 8 of which make a poppy-seed (or as some scholars hold, no poppy-seed at all but the egg of a louse)? And what exactly is meant by rabbit-, ram- and ox-grains? The size of a particle kicked up by each? But the tower of Babel suffered similar growing pains. The names in their variety and euphony stir up a magic that digits clicked off on strings knotted with zeros cannot.

A second moral lies in a remark of the Buddha's: "No being knows this counting except for me or someone who, like me, has reached his last existence, living outside of his house... this is the end of calculation. Beyond it is the incalculable." In other words, numbers cannot exceed the number of things there are: so that for the teller and the audience of this tale, numbers still are attached to objects - and more important for our story, there is no notion yet of a full place-holder system (if there were, we would see that numeration can't end, since we could make still greater numbers just by multiplying any candidate for the last by ten).

The third moral is the most significant for our tale, and it is that the Greek influence on Indian culture of this time is clear. The appearance of poppy-seeds in Archimedes' sequence and here just can't be a coincidence; but if you dismiss it, how dismiss the structural similarities between the two accounts? In fact, if you look anywhere in Indian astrology, astronomy or mathematics you'll see traces of Greek forebears: Hindu names of the zodiacal signs and various astronomical terms are Greek loan-words ('kendra' from kentron, center, for example, and 'lipta' for minute from lepton); and they wrote their fractions in the same peculiar way the Greeks did, without a separation line. Yet once again, the best evidence is structural, such as their theory of planetary motion, in texts from 400 A. D., being the Greek epicyclic one. And look to error for the surest tell-tale of truth: the ratio of longest to shortest day is given as

3:2 in early Hindu astronomy - a ratio wholly wrong for all but the most northern latitudes of India, but correct for Babylon, and adopted by the Greeks. You will find a grudging acknowledgement of the Greek source of Indian astronomy and its accompanying mathematics in the *Surya Siddhanta*, apparently delivered by the Sun to a gentleman named Maya Asura in 2,163,102 B. C. The Sun instructs him to "go to Romaka-city, your own residence. There, reincarnated as a barbarian (thanks to a curse of Brahma), I will impart the science of astronomy to you." Romaka: that is, Roman, meaning the Greeks of the Roman or Byzantine Empire; and barbarian: the Greeks again, who "indeed are foreigners," as the astronomer Varahamihira wrote around 550 A. D., "but with them astronomy is in a flourishing state."

It wouldn't surprise you, then, were the reincarnation of zero in India to be in the guise of the hollow circle we know from Greek astronomical papyri. The people of Gwalior - some 250 miles south of Delhi - wanted to give a garden to the temple of Vishnu there, from which fifty garlands of flowers could be taken each day - a lovely thought. They had the details of this gift inscribed on a stone tablet, dated Samvat 933 (876 A. D.), which shows that the garden measured 187 by 270 hastas. 270 is written 27°, and 50: 4°. This is the first indubitable written appearance of the symbol in India. Documents on copper plates, with the same small o in them, dated as far back as the sixth century A. D., abound - but so do forgeries, since the eleventh century seems to have been a particularly auspicious time for regaining lost endowments and acquiring fresh ones, through a little creative reburnishing of the past. You won't lack for people. however, who find these copper plates authentic, and wrangle with those who, they say, are just out to have the Greeks triumph over all comers.

The gods who watch over scholarship must have been off settling their domestic squabbles during these disputes, which are paved wall to wall with fallacies of negative, presumptive and possible proofs, fallacies pragmatic and fallacies aesthetic. Perhaps the world would be better and the past more attractive if some dead Indians had devised the hollow circle for zero rather than some dead Greeks (though I can't see why, especially since the concept matters more than its marker, and the concept, as we've seen, goes back to some dead Babylonians). It does strike me, however, that burdening actual Indian achievements with others' goods ends up diminishing them, and that it is a loss to replace a story rich in the accidents and ambiguities of time with an uplifting tale. Had the Indians invented a symbol for zero which was, say, a tattooed man in a necklace with his head thrown back, who would question their originality? As it is, they seem by the end of the ninth century to have long since had access to, and made good use of, Greek writings stuffed with the same symbol for zero that now took root among them.

We can try pushing back the beginnings of zero in India before 876, if you are willing to strain your eyes to make out dim figures in a bright haze. Why trouble to do this? Because every story, like every dream, has a deep point, where all that is said sounds oracular, all that is seen, an omen. Interpretations seethe around these images like froth in a cauldron. This deep point for us is the cleft between the ancient world around the Mediterranean and the ancient world of India.

There in the City of Flowers, not far from the ruined palace once built by genii, stands the astronomer Aryabhata, around 500 A. D. - but some say there were two Aryabhatas, with opposite reputations - or possibly even a shadowy third. Being an astronomer, his name - their names - should mean "learned man", as it would with two t's; having only one turns him improbably into a mercenary. Some who pursue these shadows claim that he wrote two books with contradictory statements; others, that he confined his contradictions to one - while still others say that the text of his that survived is hopelessly doctored. Is his peculiar mixture of pearl shells and sour dates (as the Arab historian Albiruni put it a thousand years ago) the product of careful observations and careless borrowings?

Whatever the case, Aryabhata wanted a concise way to store (not calculate with) large numbers, and hit on a strange scheme. If we hadn't yet our positional notation, where the 8 in 9871 means 800 because it stands in the hundreds *place,* we might have come up with writing it this way: 9T8H7Te1, where T stands for "thousand", H for "hundred" and Te for "ten" (in fact, this is how we usually pronounce our numbers, and how monetary amounts have been expressed: £3.4s.2d). Aryabhata did something of this sort, only one degree more abstract.

He made up nonsense words whose syllables stood for digits in places, the digits being given by consonants, the places by the 9 vowels in Sanskrit. Since the first three vowels are a, i and u, if you wanted to write 386 in his system (his numbers go from right to left, so 6, then 8, then 3) you would want the sixth consonant, c, followed by a (showing that c was in the units place), the eighth consonant j followed by i, then the third consonant g followed by u: CAJIGU. The problem is that this system gives only nine possible places, and being an astronomer, he had need of many more. His baroque solution was to double his system to 18 places by using the same 9 vowels twice each: a, a, i, i, u, u and so on; and breaking the consonants up into two groups, using those from the first for the odd numbered places, those from the second for the even. So he would actually have written 386 this way: CASAGI (c being the sixth consonant of the first group, s in effect the eighth of the second group, g the third of the first group). When next you are tempted to think that there aren't different minds but only Mind, remember Aryabhata.

There is clearly no zero in this system - but interestingly enough, in explaining it Aryabhata says: "The nine vowels are to be used in two nines of places" - and his word for "place" is kha. *This kha later becomes one of the commonest Indian words for zero.* It is as if we had here a slow-motion picture of an idea evolving: the shift from a "named" to a purely positional notation, from an empty place where a digit can lodge to "the empty number": a number in its own right, that nudged other numbers along into their places.

Who next can we make out in the haze, and what is that haze itself? Its particles are words, colliding with each other and diffracting the light of ideas: because once a name like 'kha' describes some aspect of zero, others will condense, until what zero *is* lies entangled in what it *does,* and what it resembles. Looking through the broken light, there - fifty years after Aryabhata, in Ujjain, stands Varahamihira, whom we met briefly, praising the Greeks for their astronomy. He hadn't a symbol for zero either, but a flurry of names for it: Aryabhata's 'kha', words for spaces, like ambara (sky) and akasa (atmosphere) - and 'sunya', usually translated "empty" - which soon became zero's commonest name. Did these derive from the yet earlier texts he mentions, at least some of which may have been those he praised, of Greek origin?

And there too in Ujjain, that center of science not very far from Gwalior, a hundred years later is Brahmagupta: Aryabhata's severest critic (and, by turns, fervent admirer: had you expected less in this play of shadows?). He too writes about zero as "sunya". Is "empty" meant to highlight the receptive nature of Aryabhata's "places"? Whatever the intention, notice how, as a substantive adjective, it brings zero that much closer to numbers, which lightly straddle the gap between noun and adjective; notice too how well it harmonizes with the once and future symbol, an empty circle.

Move two hundred years forward now to 830 A. D.; and 700 miles south to Mysore; and cross from one religion to another: from Hindus to Jains. Here is Mahavira, whose book, the *Ganita-Sara-Sagraha*, is meant to update and correct Brahmagupta's. He deals extensively with zero but still has no symbol for it - nor does he call it "sunya" but returns to "kha". Perhaps this accords with his zeal to revise, as well as with his remoteness from Brahmagupta (were we to eavesdrop on a conversation of George Washington's, we might have trouble with "hoy" and find "hoyden" quaint). But why should he have gone past Varahamihira in the number of synonyms for zero, drawn from kinds and qualities of sky and space: depth, firmament, the endless, thunderous names like jaladharapatha and gentle names like div, and Vishnu's footprint and middle air, some twelve near-synonyms for sky in all?

Was it because he somehow thought of zero as itself different in different contexts? This reminds me of the linguist's appealing point that at first we see and name as distinct what come to be recognized later as parts of a whole - which is why our eldest verbs are so irregular, since "they are" and "she is" seem very different from "I am". Or was it that the Indians, like the Greeks, tended to equate wisdom, knowledge and memory, so that important matters such as mathematics were written in the memorable form of verse. That meant having to have an army of alternatives on hand to fit the varying demands of rhythm (Mahavira also had troops at the ready for each digit). Certainly this extracting of sounds from shapes, and the storing of them in memory's figureless mansions, may have hastened on the abstracting so congenial to mathematics.

Of course this doesn't explain his having more than one term with the same scansion (nabhas, cloud, and viyat, atmosphere) - but verse-form may have inspired poetic expression. Could there also have been some impulse to conceal jewels of knowledge in brown-paper parcels? I don't know what a casual reader would have made of such phrases as "...the sky becomes the same as what is added to it."

Or could Mahavira have wanted metaphor to carry his mathematics across to other realms? We have to consider too that the Salutation of his book concludes: "May the rule of that sovereign lord of the Jinas prosper, who has destroyed the position of single conclusions and propounds the logic of the syadvada." The syadvada, his English translator explains, is an argument that the world of appearances may or may not be real, or both may and may not be real - or may be indescribable; or may be real and indescribable, or unreal and indescribable; or in the end may be real and unreal and indescribable.

Which of these combinations best fits zero? Which best describes its status at that time, and our knowledge of it? The more names it had, you would think, the less of a proper number it was - still part of discursive language rather than mathematics. But what if there had long since been a sign for it too, as some claim? They call to their aid a retort from the Euphrates, where in 662 the Syrian Bishop Severus Sebokht exclaimed that the Greeks hadn't the monopoly on science but were merely pupils of the Chaldeans of Babylon; that not they but the Syrians invented astronomy; and that besides, the discoveries of the Hindus were more ingenious than those of the Greeks, with methods of computation that surpass description. "I wish only to say," he adds, "that this computation is done by means of nine signs." *Nine?* Doesn't this testimony in fact add to the evidence that the Hindu zero still awaited its incarnation as a sign, living the while, if not in words, then in that intermediate world of spaces between numbers?

Once again, while having a symbol for zero matters, having the notion matters more, and whether this came from the Babylonians directly or through the Greeks, what is hanging in the balance here in India is the character this notion will take: will it be the idea of the absence of any number - or the idea of a number *for* such absence? Is it to be the mark of the empty, or the empty mark? The first keeps it estranged from numbers, merely part of the landscape through which they move; the second puts it on a par with them.

High romance has gathered around this pregnant moment. A hundred years ago people said such things as: "The philosophy and religion of the Hindus uniquely fitted them for the invention of zero", and that their inventing a symbol for zero was like making Nirvana dynamic. In that Bible of our grandparents' generation, *The Decline of the West*, Oswald Spengler wrote that zero was "that refined creation of a wonderful abstractive power which, for the Indian soul that conceived it as a base for a positional numeration, was nothing more nor less than the key to the meaning of existence." The Greek soul, he informs us, is sensual and so could never have come up with this key: it takes a Brahmanic soul to perceive numbers as self-evident.

We dismiss Spengler's dicta now for the very good reasons that much of what he based them on he simply got wrong; and that the language of Race-Ideals, Destiny and The Faustian Soul, so exciting in the gloom of 1918, was spoken by demonic voices twenty years later. But we also dismiss them because, in our cautious times, we distrust generalizations, and would rather forego what unity they might bring to the little chaos of things than risk being taken in by glib conclusions. We romanticize ourselves as hard-headed. None now, surveying from afar the culture of

India, would risk saying with Spengler: "Only this spiritu-ality could originate the grand conception of nothingness as a true number."

Instead of imagining the Hindus as deriving the hollow circle for zero from the meaning of existence (or is it the other way round?), some scholars, defending its Indian ori-gin, make a fascinating Finneganbeginagain argument based on the old Brahmi numeral for 10 - which was ⟨α, perhaps, or α⟩ (in the almost illegible inscriptions of a cave on the Naneghat Hill near Poona), and ⟨α and ο⟩ from the first or second century A. D. in the sacred caves at Nasik.

What if someone at last thought of 10 (this argument runs) as the start of the second tier of counting, so that the first tier needed to begin not with 1 (which would corre-spond to 11) but with the analogue of 10: so with ⟨α⟩ shorn of its spurs, hence simply 0? In questionable support of this claim they point out that early European arithmetics always list 0 after 9, on the Arab model. This conjecture seems to require the Brahmi 1 to look like ⟨< or -(or =⟩, whereas in fact it was written as you would expect: I or __ . Unfortunately too the same inscriptions use 0 for 20, while 0 is used for 1 in the Sarada numerals from Kashmir, of uncertain antiquity; and elsewhere in India at this time, 10 appears in the varied guises of ⟨7 . У . 6 , ∠ and π⟩.

Even if we accept this connection of 0 and , Greek prece-dents crop up again. As their first letter, alpha (written a) stood for 1 in the Athenian system you saw on pg. 15. Per-haps as a visual pun combining the letter for 1 with the image of a single stone, the Greeks from early on often rep-resented their shaped numerals not with solid dots as on pg. 20 but with alphas; so that the tetractys, for example - the figurate 10 from which the Pythagoreans saw all of nature flowing - was written:

α
α α
α α α
α α α α

Could this talisman in turn have been condensed to a single a by the time it reached India?

The historian of mathematics Karl Lang-Kirnberg plays a variation on this that ends up taking the laurel away from the Indians, the Greeks and even the Babylonians and fitting it snugly on Sumerian brows, back in 3000 B. C. You remember that before they wrote with a stylus, the Sumerians made their marks in clay with a reed - and their symbol for 10 was this reed's unslanted impress: O. By some legerdemain which Lang-Kirnberg hurries us past, it then (or thus?) made the number to its left ten times greater, and so by itself came to stand for zero. "O could not multiply a number by 10", he asserts, "if it had not originally itself had the number-value 10." But where or why this source of zero had been hiding in the intervening millennia, or why it chose to step out just where and when it did - here Lang-Kirnberg falls as discreetly silent as did Scheherazade when she saw the dawn appear.

Do you begin to feel that every ecological niche in the world of zero's possible origins has been filled? Would you care to consider too that in Sanskrit the sign for an omitted word or syllable is - of course - a little ° ? And that among the Tartars a superfluous part of a text was enclosed in an oval? Or that by 1150 the Indian mathematician Bhaskara indicated which of two numbers was to be subtracted from the other by placing a small circle above it? The circular rings are expanding and pooling new circles everywhere.

Or have you concluded that there aren't very many symbols, after all, which can easily be written and read, but that there are a great many ideas and operations needing symbols, and we should count ourselves lucky that context helps us to figure out whether we're reading about degrees, monads, 70 myriads, subtraction, obols, words superfluous or omitted, stones or the absence of stones, 70 or 1 or 20 or 10 or nothing at all.

The spate of conjectures about zero's genesis wells up from the nature of historical thinking itself. We try to retrieve what happened in a remote and sparsely documented past. Clues are thin on the ground but our minds are as ingenious as they are restless, and snatch at anything flinty to strike a spark with, in order to light up the murk. You hope that the next bit of evidence or proposal won't just lengthen the list but begin to make some connections in it. Listen again for a moment to the Bishop of Nisibis, Severus Sebokht, and your wish may be granted.

The Hindus had methods of computation, he said, that surpass description. If you wonder how that could have been, with such a mixture of signs and words and too many synonyms, so do I. The answer has to be that they didn't compute with these words, but like the Greeks used a counting board for calculation and all of those verbal systems only for storing up the results. What do we know about their board? Something wonderful: it was lightly covered with sand! In fact, the word for "higher computations" is dhuli-kharma, 'sand-work'. So it is in the dust or sand of India, Horatio, that we find the evidence we sought to support our conjecture that 0 arose from the depression left by a circular counter's removal.

5

DUST

But why should the Indians have dusted their counting boards with sand? What seems most plausible to me is that the sand acted as a memory: you can see the traces of the numbers you began with after the calculation is over, and so check it. Without sand, for example, 47 - 34 on a counting board looks like this:

Before   After

•   •   •   •
•   •   •
•   •   •
•   •
     •
     •
     •

No way to catch a deliberate or hasty mistake. But with sand you would have:

Before   After

•   •   •   •
•   •   o   •
•   •   o   •
•   •   o   o
     •   o
     •   o
     •   o

It could be, however, that counting boards had begun as furrows drawn in the sand, or were occasionally made there impromptu, and the sand was retained by tradition once the boards were fixed in wood or stone.

If this is so, a gentle tributary of conjecture flows from it. For those furrows would soon enough be blurred or wiped smooth by the movement of hands or pebbles across them, and this alone might have been enough to think of writing numerals you could erase in computations without disturbing the columns, and thus a royal road to positional notation. Since we are back to wondering what song the Sirens sang, let's add, by way of evidence, Gerbert's apices. These were counters made of horn for the counting board, devised by the monk Gerbert around 967 - before he became Pope Sylvester II. Their name came from the Latin *apex,* as if the counters were tips of cones - perhaps from an earlier variation of piling up the counters. What was peculiar about them was that each bore a different digit, so that instead of showing 47 as above, you would simply put down

Since the numerals he used were the West Arabic ones, in which the two, three and seven were written ᴎ, ⱳ and ⱴ, it is pleasing to see our own two three and seven as Gerbert's apices carelessly rotated: from ᴎ to 2 // ⱳ to 3 /and ⱴ to 7. This brings us to the conjecture that these apices were an intermediate stage between plunking down the right number of single counters and writing numerals. Of course we would then have to claim that Gerbert had only rediscovered them, since such intermediates would have to have arisen long before.

One last rivulet branching off from this conjecture: Gerbert - or his disciples - had a zero sign on one of his apices,
written like this: ⊘ . Its name, he says, is 'sipos'. Is this a
corruption of the Greek word for pebble, ψῆφος, psephos?
If so, it shows once again the intimate connection or confusion among counters and places and zero, that absence of
counters - like Aryabhata's 'kha'. It may also explain why
Gerbert was accused of having criminal intercourse with
evil spirits, since dealing with mathematics is bad enough,
but letting nothingness loose in what passed for a civilized
world just wouldn't do.

Ancestral shades of these sand-strewn Indian boards flit
in the half-light. The learned dust that Cicero speaks of,
pulvis eruditus, was the sand in which mathematicians
drew their diagrams - but that doen't make their surface a
counting board; so that when he dismissively says: "You
have never learnt mathematics", numquam eruditum illum
pulverem attigistis (literally, never touched learned dust),
the reference is probably to geometrical figures or those figurate numbers of pg. 19. The counting board sprinkled
with green sand and blue sand that Remigius of Auxerre
described in 900 A. D. sounds like something one would
dearly love to own - but since he says that figures were
drawn on it with a pointer (radius), it belongs to the same
tradition, which also produced the wax tablets that
Horace's schoolboy hung over his arm, and the slates that
long after screeched in village schoolrooms. But the Roman
board on which pebbles *(calculi)* were placed in columns
was sanded too, and the Greek word for counting board,
'abacus', αβαξ (abax), may well have come not from 'leg-
less table' but the Semitic abq, "dust".

An increased likelihood, then, that the Greek hollow circle for zero came from the impression of stones removed
from a sand-covered counting board. If there were words
for the little pleats between the possible and the probable

we could choose one of them here - some junior member of the family that gave us would, should and could. Without them, let us fold up the corner of this conjecture lest we find ourselves putting words in ancient mouths as some anonymous scholar did in the eleventh century, reading his Boethius. Why not add to that fifth century Roman's already encyclopaedic works a book on geometry, he must have asked himself, complete with the numerals zero through nine that had come out of India so recently, via the Arabs; along with a statement that the counting board was in fact the Mensa Pythagorica, Pythagoras' Table, decorated with Pythagoras' Arches - so that subsequent readers would clamp esoteric traditions and arcane gossip together with the authority of the man who had written in prison *The Consolations of Philosophy*, to deduce that Pythagoras had brought the board back to Greece from his eastern wanderings. And so what we wish had happened becomes part of what happened too.

We fold this corner only to open another that dust spills out of: for here in Moorish Spain, as early as 950 A. D., are figures the Arabs called huruf al gobar: "dust numerals". What are they, where did they come from, why this peculiar name? They are the numbers 1 through 9, without a zero, although their name is thought to have come from those dusty Indian boards by way not of scholars but of merchants in their travels. What matters for our story - as if halfway through a mystery fresh suspects suddenly appeared - is that these dust numerals have themselves an odd sort of dust around them, showers of dots that indicate their place-value. No dots above a numeral meant it stood for units; one for tens; two for hundreds - and so on; so that (using our numerals rather than theirs) 83 would mean 8030, but 83 would be 8003.

Dots, solid if minute dots, acting almost as zeroes do in positional notation: is this a congested variant on the hollow circle, or traces of a wholly different genetic strain in the family whose fortunes we have been following? Is this the n'er-do-well uncle come to assert his superior claims to a fortune? For if you now look with dots in mind you'll find them everywhere, acting as zeroes or in ways that recall the void, the missing, the absent, the indefinite, the invisible.

One of history's appealing paradoxes is that we have to look ahead to see behind: look at the scattered effects to deduce causes, at likenesses among descendants to know their ancestors' features. So here. Let's follow the swarm of dots we find in writings from a host of languages, across great spans of time, and on topics mathematical and otherwise, to see if they will lead us to the hive.

As sounds dropped out of Greek and Latin, Sanskrit and other tongues, at least as early as the first century B. C. they became diacritical marks: nasals and Gs and Hs that lost their identity, fell silent, and lived from then on largely as dots or strokes or curls on other letters - what one linguist speaks of as zeros, letters expressing nothing, hence able to serve as signals to modify more vigorous sounds. Is this an aural parallel to zero's role as a shifter in value? In Hebrew when vowels appear they often do so as dots above or below other letters, serving as aids for beginners or to prevent ambiguities (or, by their absence, promote them, as in one form of interpretation where words were read as if they were differently vocalised and a different meaning was therefore concealed within the original). It was in this pointillist atmosphere that the antecedents of the gubar numerals arose.

Among these ornamental dots are the rare, mysterious ones - only 15 in all - over and sometimes under words or letters in the Torah and Prophets. Evidence carries them back to the second century A. D., but what they signify is the stuff of scholarly dispute. Do they undermine the meaning of the word they fall on (the way we raise our eyebrows at one another in print via quotation-marks)? The commentator Rashi says of one instance that it may have been intended to make it seem that the word had not been written. Rather like assigning it the value zero, or taking it off the board. Another kind of dot could make the difference between life and death, since it was a capital offense to write on the Sabbath. Writing was interpreted as setting down more than one letter. But what if you used a single letter to abbreviate a word? Opinion was divided over whether this would be criminal or not, and brought in its wake the question of how to know if an abbreviation had been meant. Rashi remarked that a dot may have distinguished an ordinary letter from an abbreviatory one. A single way to show so many shades of absence - even car horns are more expressive.

When we go back to the Indian context we find a dot used to stand for a pledge to complete an unfulfilled task, but also to indicate a lacuna in an inscription or a manuscript: 'sunyabindu', "the dot marking a blank" (think of our using three dots to leave to the reader the obvious completion of our thought, or as much as to say: it is all too hard to put into words, but you know what I mean). In Sanskrit, when the bindu (here called kha) represents the nasal n, it also carries great mystical significance, connecting it to the god Siva.

Given these phonetic and syntactical and semantic zero-dots, should we take the hints from the Arabic gubar numerals that there were Indian mathematical zero-dots too? A more definite if slightly skew pointer lies in *The Book of Lists (Kitab al-Fihrist)* compiled by Ibn Abi Ya'qub al-Nadim in 987. He alone describes the Indians as using *subscripted* dots in the same way (one dot for tens, etc.), and under numerals older than and different from the gubar ones. If such dots - over or under - preceded the hollow circle, might they explain why it was always written half the size of the other figures? Or does its minuteness just testify that, like the dots, the zero-symbol wasn't on the same level as letters and numbers proper, but acted as a modifier or as a separator, the way our periods and commas, colons and semi-colons do? If you want more evidence that circle followed dot among the Indians, look at the circles Bhaskara, in 1150, put over a number that was to be subtracted: this had been a dot five centuries before, in Brahmagupta. Certainly for Bhaskara dot and circle lived comfortably together: as one of his commentators wrote, "...the place, when none of the nine numbers belongs to it, is shown by a blank (sunya), which, to obviate mistake, is denoted by a dot or small circle."

The important question, however, is: how far back does the dot as zero go? Such questions always seem straightforward until you try answering them. References in Arabic texts to Indian work are affected by the fact that in Arabic the zero was always written as a dot, since for them the hollow circle was reserved for 5 (the list of numbers that *didn't* have '0' as their symbol is rapidly shrinking). Much has been made of an arithmetic written on birch-bark that a farmer dug up, mutilated, in 1881 in Bakhshali, on India's northwestern frontier. It is rich in dots for zero - here too called sunya - but how old is it? It was once thought to date from the third or even second century, but that belief is disintegrating faster than the birch-barks themselves, and current opinion traces the passages with zero in them to about the seventh century A. D.

You'll find firmer ground, and a striking image, in a celebrated book from around 620, the *Vasavadatta* of Subandhu. In it he says that the stars dot the sky like zeros "because of the nullity of metempsychosis", the Creator reckoning the sum total on the ink-blue sky with a bit of the moon for chalk. Receding one more century, do you remember Varahamihira and his synonyms for zero? One was 'bindu', dot (though he used only names, not symbols) - and this seems as far back as we can safely go. Before this the landscape is nearly empty of historical landmarks, blending into myth.

Nearly empty: but not quite. In 270 A. D. someone named Sphujidhvaja wrote the *Yavanajataka*, "The Horoscopy of the Greeks", a translation into verse of Sanskrit prose from 150 A. D. The Greek original behind it was almost certainly from Alexandria. In adjacent sentences of the text, restored in 1978, we find the number 60 mentioned twice: first as "sat binduyutani", then as "sat khayutani": that is, "6 with 0", i.e., 60. The word for zero, as you see, is bindu the first time and kha the second. 150 A. D.: the time of Ptolemy's *Almagest*. The time when a marker of clay - a ball or a bead - was spoken of (as Solon and Polybius once did) as having different values in different positions. What earlier reference could you ask for - but more to the point, what better evidence that the hollow circle of kha and the solid dot of bindu came to India from Greece?

6

INTO THE UNKNOWN

How clear it all seems now, how certain... until you reflect that we've tried to bridge a chasm on the slenderest threads of evidence, as steel cables were once drawn over Niagara Falls by attaching strings and then wires of ever greater diameter to the tail of a kite flown across it. Will we be able to hang a great roadway from ours? No sooner was the translation of Sphudjidhvaja's text published than the restoration of the two sentences that seemed to involve words for zero was called into question. The 'kha' and 'bindu' one scholar had called forth from the Plain of Ghosts were sent howling back by another, and there for all I know they consort still with half-likely creatures, cooking their thin broths in those pots that Irish folktales tell us were made by first finding complete emptiness, then pouring molten iron over it.

Shall we scribble across our canvas, as da Vinci did again and again in old age, "Di mi se mai fu fatta alcuna cosa": Tell me if ever *anything* was finished? Shall we let the question fittingly trail off with a series of dots? No: once more into the breech. 'Bindu', after all, which means a breech, also means "the sudden development of a secondary incident, expanding like a drop of oil in water to further the plot". Here we may have just such a development: something which could tell us - not whether the Indians came up with the dot or circle for zero - but much more significantly, how they thought about this zero once they had it. Remarkably enough, the dot was used by them not only for zero but for the unknown, the way we use x. So in the Bakhshali manuscript a problem we would pose as: "What is the whole number found when 27/8 is multiplied by 32?", or:

$$x = \underline{27} \cdot 32$$
$$8$$

(so x = 108), they write this way:

•	3	3	3	32	=	108
1	2	2	2			

(the 1 under the dot shows that the unknown is a whole number).

This may not strike you as being much of an unknown, since the question is simply: what is

$$\underline{27} \cdot 32 \ ?$$
$$8$$

But another problem in the Bakhshali arithmetic will bring back the familiar lurch in the stomach from your schooldays:

B gives twice as much as A, C three times as much as B, D four times as much as C. Together they give 132. How much does A give?

The method shown for solving this problem is ingenious:

Take 1 for the unknown (sunya). Then A = 1, B = 2, C = 6 and D = 24. Their sum is 33. Divide 132 by 33 and the answer, 4, is what A actually gave.

(We would say: let x stand for what A gave. Then B gives 2x, C gives 6x and D gives 24x. so x + 2x + 6x + 24x = 132, or 33x = 132. Hence x = 132/33 = 4.)

We can't pin down how early the use of a dot for the unknown was. Brahmagupta, in 630 A. D., called his variable "as much as", 'yavat tavat', abbreviated 'ya' (and when he needed more than one variable, as we use x, y and z, he gave them color names: black, blue, yellow, white, red, abbreviated: ca, ni, pi, pa, lo). But a time came when an Indian mathematician, leaning out of legend, saw no problem in calling both nothing and something 'sunya', and the usage stuck. How could this be? The American logician Willard van Orman Quine once pointed out that 'nothing' and 'something' are false substantives, behaving like nouns grammatically but not logically. Nouns name things, he wrote, and a thing cannot be both red and not red, for example. But "something is red" and "something is not red" are both true (the story is told of Quine that when a pianist playing Mozart apologized for striking a wrong note, Quine assured him that he had just played something else perfectly).

What can be nothing one moment and something the next, yet disappears in the presence of anything? This sounds like one of those conundrums dear to nervous people at parties, but is in fact the puzzle at the heart of the Indian 'sunya'. The answer lies in our always having mistranslated this word by 'void' or 'empty'. For the Hindus there is no unqualified nothingness. In the same spirit as our Law of the Conservation of Matter, substance for them cannot disappear but only change its form or nature: this fullness - brahman - pervades the universe, and can no more increase or decrease than can the 'absolute element' that plays a similar role in Buddhism, which is empty (sunya) only of the accidental.

Or put it this way: it is as if there were a layer behind appearances that had no qualities, but took on the character of its surroundings, accomodating itself to our interpretations, as ambergris acquires and retains fugitive fragrances, giving us perfume. 'Sunya' isn't so much vacancy, then, as receptivity, a womb-like hollow ready to swell - and indeed it comes from the root svi, meaning swelling. Its companion 'kha' derives from the verb 'to dig', and so carries the sense of 'hole': something to be filled. And two names for brahman - akasa and purna - appear also as names for zero: the first in Varahamihira and Mahavira, the second, later, in Bhaskara.

This is the zero of the counting board: a column already there, but with no counters yet in it. This is the zero of the place-holder notation, having no value itself but giving value by its presence to other numerals. These same qualities belong to the variable, the unknown: a potential which the different circumstances of the equations it lies in will differently realize. The background shift is from counters taking their value from being in different places, to a single, receptive place whose circumstances will reveal its hidden value. So one sets at the Passover table an empty place for Elijah. He may come in the form of a beggar or to announce the end of days. Like Lord Krishna invoked by the milkmaid's song in E. M. Forster's *Passage to India*, he may not come at all. "I say to Him, Come, come, come, come, come, come. He neglects to come."

This sense of 'sunya' chimes with Mahavira's logic, where the emphasis falls on the ambiguous and indeterminate nature of appearances (linking them back to their source in the potential); and it helps us to understand another salutation - this one at the beginning of Bhaskara's book on mathematics, the *Vija-Ganita*: "I revere", he says, "the unapparent primary matter... for it is the sole element of all which is apparent... the arithmetic of known quantity... is founded on that of unknown quantity; and... questions to be solved can hardly be understood by any, and not at all by such as have dull apprehensions, without the application of unknown quantity..."

Haven't our dots all funnelled back to India? Were zero and the variable not truly born here, twin offspring of sunya and what seems the singularly Indian understanding of vacancy as receptive? Perhaps Spengler was right in the end, that only Indian culture was attuned to produce these notions.

But like an hour-glass, the funnel opens out again and the dots stream down to ancient Greece. The problem with using letters for numbers is that you need to distinguish them from words by some sort of marks. The line you saw on pg. 15 which the Greeks drew over such clusters was often broken up into short dashes or even little loops on each: so $\overline{X}$ or $\overset{\smile}{X}$ for 600. Some writers used one or more dots fore and aft instead: ⁙ for 600, :t i h: or t i h for 318. The standard way they increased a number's value by 1000 was to put a little mark down on its left-hand side: β was 2 but ,β was 2000 (at times this stroke grew out of the letter or floated above it: A and A were each 1000). The penchant for these strokes knew no bounds. Fractions were sometimes indicated by putting the stroke to the right and above, so that γ was 3 but γ' (or at times γ or even γ'') was 1/3. Archimedes wrote 10/71 this way: ι οα' (ι was the numerator, 10; ο was 70 and a 1, so οα was 71, and the stroke showed that this number was in the denominator). You marvel yet again that he could carry out the involved calculations he did.

That triad of great Alexandrian mathematicians of the second and third centuries A. D. - Heron, Pappus and Diophantus - were well situated to have their works influence India, and their notations were even nearer our concern. Diophantus, you recall, separated his myriads from his units by putting the sign M̃ between them. But he and Pappus at times just put a dot there, so that β·οδ, for example, stood for 20,074. In effect, then, this dot multiplied what was to its left by 10,000, behaving like a cousin of zero. Heron, who invented the first steam engine and wrote on the art of constructing automata, multiplied his units 10,000 times by putting two dots over them: α was 1 but α̈ was 10,000. In another old Greek notation, this series continues, each new dot multiplying the value by 100. Did the Hebrew conventions of placing two dots above an alphabetic numeral to increase its value by 1000, and distinguishing such numbers from words by small superscripted accents, derive from Greek practice, or vice versa?

Very far back, then, the Greeks used dots in the way we saw them used later in the gubar numerals. I've argued that Indian mathematicians had the same sign for zero and the unknown because they thought of each as an unfilled container. If I'm to carry this argument back to the Greeks, I must show not only that they too used a dot for zero, but that they operated with an unknown; and used the same symbol for it and for zero; and had reasons (similar to the Hindus or not) for doing so.

In fact the unknown was - what should we say, discovered or invented? - far back in Babylon, popping up now and then in little puzzles, and in a not so little way around the time of Plato, when a Pythagorean named Thymaridas saw how to solve certain simultaneous equations in several unknowns. This rule is called his Flower. Who can say why we hardly hear of the unknown again until Alexandrian times (is this an instance of Pythagorean secrecy?) - but when we do, it has a blossom of its own for us.

Diophantus calls it 'arithmos' (αριθμος), 'number', and defines it as containing "an indeterminate multitude of units" (we see a narrow flow of continuity from Thymaridas, six hundred years earlier, in their both using the word 'indeterminate' - αοριστος). And his symbol for it? Occasionally ५ (recall from pg. 16 that ५ was the sign for zero in a few late Byzantine texts), but usually contracted from its first or last two letters to a shape like ς , at times ς° - and sometimes just ° ! Not only Diophantus but the Greeks of his day used this symbol (there is a form of it, for example, in a papyrus from 154 A. D.). Did the small circle here, like that for degree on pg. 16, bubble up and bob along to India, carried on the tide of these Alexandrian works? Certainly the presence of Diophantus is palpable there: that clever way of solving a problem by taking its unknown to be 1, which you saw in the Bhakshali manuscript, occurs earlier in Diophantus in a problem yet more involved.

And the pregnant void? You'll find something strikingly like it in Aristotle, but with two peculiar twists. He says that it is a common belief that "void is a place where no body happens to be" - a place which is just temporarily deprived of its content. It sounds, then, as if Alexander's tutor had perfectly anticipated the Hindu sunya. The first twist (which tells you more about Aristotle than sunya) is that he seems himself to have come up with this definition and foisted it on his predecessors. Very well; but even if it wasn't a "common belief" among the Greeks, it is so put in Aristotle's *Physics*, and therefore resounded as such down the centuries and across the contemplative world. The second twist, however, makes a Gordian knot of the void, for Aristotle no sooner defines it than he shows that it doesn't exist! Void is a place where body may be; but with eternal things (and the elements that make up body are eternal), there is no difference, he says, between possible and actual Being - so all places are occupied. And besides, he only brought up the void in order to show that he had no need of it in explaining motion. In fact, its existence would be an obstacle for his theories: so he dismissed it. Can a definition of something deprived of reality give rise to a lively concept?

Fortunately we don't have to answer this question, because resonant as Aristotle's *Physics* was, it echoed no louder than a whisper compared to Plato's *Timaeus* until well into the twelfth century. This late dialogue of his, written somewhere around 350 B. C., has been the wonder of his readers ever since. What does it mean, and how does it mean? In that first millenneum after he wrote it, it was taken as a work of great mystical vision that needed intitiation into arcane secrets if its message was to be revealed.

Almost at its very middle lies the revelation for us. Timaeus - perhaps a Pythagorean astronomer and mathematician, perhaps a pure invention of Plato's - has been presenting a detailed cosmogony, and finds himself compelled to start again, the discourse having deepened. Before, he had spoken of Being and Becoming. Now he realizes that there is a third factor involved in the creation of the cosmos:

> ...the argument compels us to attempt to bring
> to light and describe a form difficult and ob-
> scure. What nature must we, then, conceive it
> to possess and what part does it play? This,
> more than anything else: that it is the Receptacle -
> as it were, the nurse - of all Becoming.

He finds it hard to explain the nature of this Receptacle - as would anyone who was trying, for the first time, to come to grips with the notion 'variable' or 'unknown', as I think Plato, in the persona of Timaeus, is here trying to do. He speaks of it as having a nature which receives all bodies:

> ...it is always receiving all things, and never
> in any way whatsoever takes on any character that

is like any of the things that enter it: by nature
it is there as a matrix for everything, changed and
diversified by the things that enter it, and on their
account it *appears* to have different qualities at
different times...

Still struggling to express this notion, he compares it to a
mother, and continues:

Hence that which is to receive in itself all kinds
must be free from all characters; just like the
base which the makers of scented ointments
skillfully contrive to start with: they make the
liquids that are to receive the scents as odor-
less as possible... We shall not be deceived if
we call [this mother or Receptacle] a nature
invisible and characterless, all-receiving, par-
taking in some very puzzling way of the intelligible and very hard to apprehend.

Then, without warning, he suddenly identifies this receptacle with space!

...space, which is everlasting, providing a situation
for all things that come into Being, but itself
apprehended without the senses by a sort of bastard
reasoning... Let this, then, be the tale according to my
reckoning: that there are Being, Space, Becoming - three
distinct things, even before the heaven came into being.

Timaeus told us this was going to be hard. The passage
of time and the loss of context haven't helped, and Plato
may have intentionally obscured matters in order to make
us work through the ideas ourselves - or to keep them from
those unwilling to go through this initiation. Still, the shape

of this constellation shines through, and it looks very much like the one we've already seen from Aristotelian and Indian vantage-points. The word Plato uses for space (chora, χωρα) carries the sense of a container ready to be filled, like Aristotle's void and the Indian sunya; and what Plato proceeds to fill it with is - numbers! More precisely, with the various elements, thought of as made from those figurate numbers we saw on pg. 20.

We are used to thinking of the unknown in an algebraic context - that is, after all, what algebra is about: equations with numbers and unknowns and discovering what the unknowns must be. But Plato's intuition was geometric, and what we have here seems to me to be the perfect geometric analogue of algebra, with shaped numbers filtering through empty space. To accord with the indeterminate character of what is held by the nurse, he says that we shouldn't call the elements in it - air, made of octrahedra, for example, and water, of icosahedra - by fixed names, but only refer to them as "the such-and-such" (το τοιουτον). Perhaps Brahmagupta's name for the unknown, yavat tavat, "as much as", owes something to this. Plato even stresses the mathematical character of his account by having Timaeus say twice, in presenting it, "Here is how I reckon it", and "Here is how I sum it up", where the words he uses are from 'logistic', which you saw on pg. 19 was the scorned art of calculation; and psephos (ψηφος), the stone used as a counter.

To sum it up ourselves, the concerns of philosophers in classical Greece, at a time just prior to the spread of their thoughts to India, were such as made an appropriate setting for considering zero and the unknown in similar ways, so that it shouldn't surprise us that the symbols for the two intermingled.

The loosest link in our chain made of flowers and bubbles and nurses is that this symbol among the Greeks was more circle than dot, but among those in India, more dot than circle. Might the signs then after all have originally been independent, the zero notion they stood for not transplanted but grown from each culture's native stock? Here is a last line of speculation to follow which may bind them up together - keeping in mind that speculation and scholarship live like two brothers in the same house, whose divergent characters have left them on distantly formal terms with one another. It begins at the end, with Gerbert's apices. You remember that for some reason he had a counter for zero, and that on it the symbol was peculiar: **◑**. This same symbol surfaces from time to time in the middle ages, looking like the Greek letter theta, θ, and hence (you might think) called 'theca'. It was explained by Petrus of Dacia in 1291 as coming from the brand burned into the cheek or forehead of criminals - for you might need a crossbar to attach the iron to the handle; and criminals were the zeros of society: we still call our losers zeros, as do the French.

Now there was a learned man named Adelard of Bath, who early in the twelfth century left England to become a tutor in Laon; played his cithara before the Queen of France; made his way to Spain, to study in Toledo; picked up some Arabic and went to Cordoba disguised as a Mohammedan; and like many an Englishman after him, set out for the East. We catch a glimpse of him stopping to talk to an old philosopher outside Salerno about magnetism; sailing across to Greece; pushing on to Cilicia and Syria. There he is, observing that light travels faster than sound. And there he is again, badly shaken on a bridge near Antioch when the earthquake struck that razed many a city in the levant.

When he came home at length it was with a spirit touched by the Renaissance pre-dawn ("If you want to hear anything from me", he said, "give and take reason"). And he brought back with him precious manuscripts, the real treasures of the East: a treatise on alchemy thinly disguised as a text on mixing pigments (though it also contained a recipe for making toffee), works on how to build foundations under water and how rightly to spring vaulted structures. He wrote a book of his own on falconry, in the form of a dialogue with his nephew. In later life wore the green cloak and emerald ring of an astrologer, and cast horoscopes for King Stephen.

But he also brought back mathematical works ("dangerous Saracen magic", William of Malmsbury called it) which he - and later his Irish student, it seems: a certain N. O'Creat - translated from Arabic: the thirteen books of Euclid, and the astronomical tables of the great Al-Khowarizmi. In their translations we find three different symbols for zero: θ (theta), the familiar o, and   , called by him 'teca'.

"Theca" isn't a likely transformation for "theta" to undergo, and now here is "teca' to join it. Is there a more plausible explanation? There is. "Theca" in Greek means - a receptacle; and when you write it in upper-case Greek letters it looks like this: ΘΗΚΗ. Look at that first letter, theta, a dot with a circle around it. Radulph of Laon, around 1100, pictures it so, and describes it as the figure which stands for no number, whose name, he says, is 'sipos' - remember Gerbert's 'sipos' for zero, so much like the Greek 'psephos' for counter (Adelard calls it 'sipocelentis'). At this time too Rabbi ben Ezra calls it both 'sifra' and 'galgal' (the Hebrew for 'wheel') - and 'kha' also means the hole in the nave of a wheel through which the axle runs.

We may have just seen a glimmer through the dark ages from a thousand years earlier, in which our two symbols merge into one - or is it a puckish zero, leading us astray?

7

A PARADIGM SHIFTS

Histories differ from mysteries in that all of their leads are in some sense true. We know we've been playing a chancy game in trying to pin down the vagaries of zero's infancy, and even in trying to guess what shapes it put on; but like Quine's pianist, we may have been playing something else perfectly. For common to everything we've learned is this: that despite its power to extend the empire of numbers, we have yet to see zero treated as a number itself. It evolved from a punctuation mark and long kept its supernumerary character - no more a number than a comma is a letter. Even as late as the twelfth century in India, Bhaskara and his commentators ascribed the invention of the nine digits to the beneficent creator of the universe, but distinguished from these his inventing as well the place-system to make these digits serve for all numbers, along with the zero - whether dot or small circle - put in a place where no number belongs "to obviate mistake" (as you read on pg. 56). One of our commonest words for zero, "null", comes from the medieval Latin "nulla figura", 'no number', and a Frenchman, writing in the 15th century, expressed the popular view well: "Just as the rag doll wanted to be an eagle, the donkey a lion and the monkey a queen, the zero put on airs and pretended to be a digit."

On the merry-go-round of causes and effects, several factors helped to keep zero aloof from numbers. Each number pertains to specific collections of things but zero to no thing at all. Hence its easy alliance with the sign for and notion of 'variable': an alliance that then helped reinforce the distinction. And notice too that zero often comes up in the context of subtraction and negative numbers (no accident, then, that Bhaskara placed a small circle above a number to be subtracted). Any five-year-old will tell you that negative numbers aren't numbers at all, and phylogeny recapitulated ontogeny in taking its time to recognize negatives. Is this because negation is harder to picture and to grasp than affirmation, and carries zero with it into the perilous land of half-beings? Let's face it, the retrograde motion involved in subtraction makes counting, which was hard enough, thoroughly confusing, as you will know if you've ever been tricked into believing you had 11 fingers (5 on your left hand, and - counting backwards - 10, 9, 8, 7, 6 on your right, so 6 + 5 = 11). Yet without subtraction we wouldn't have this excellent riddle: four people are in a room and seven people leave it. How many must go in before the room is empty? Answer: three.

Zero's involvement with the *operations* of addition and subtraction alienated it even further from the quasi-objects that numbers were turning into. It wasn't just the aftermath of removing counters from a column, since aftermaths are still faintly things. Zero, as we saw before, was a passing condition of parts of the board: it was becoming more an action than object, more verb than noun. Mahavira put this vividly when he said that "zero *becomes* the same as what is added to it."

But Mahavira, and the Indian mathematicians we've come on through six centuries of fragmented time, did something far greater than catch a fleeting likeness of zero. They began to describe how zero behaves with other numbers, and those numbers among themselves. These descriptions took the form of laws governing their interactions. The effect of such laws would be not only to bring zero and numbers closer together but to change our understanding of numbers themselves, making an ideal country to which they - and who knows what further species and new landscapes - belonged. This was their own, and their finest, achievement.

What does it take for an immigrant to the Republic of Numbers to gain citizenship? Think of the situation with words and with ideas. New words are always frisking about us like puppies - one month people go "ballistic" and the next "postal" - but few settle in companionably over the years and fewer still reach that venerable state where we can't imagine never having been able to whistle them up, there at our bidding. And ideas, large and small: where was flower power fifty years ago - and where is it now? With what fear, fascination and loathing Freudian doctrine slowly took hold and became the canon - and how quickly it all fell apart: who now have complexes, or cathect their libidos onto father-figures?

But the Republic of Numbers is vastly more conservative than those of language or of ideas: Swiss in its reluctance to accept new members, Mafiesque in never letting them go, once sworn in. Think of irrational numbers, the guilty secret of the Pythagoreans, whose exposure shook Greek confidence to the core. Twenty-five hundred years later we can't do without them, though the sense in which they exist is debated still. And imaginaries? Mathematicians, who love high-wire acts, began thinking about the square roots of negative numbers as far back as Heron and Diophantus, but whenever these came up as solutions of equations they were called fictitious and the equations judged insoluble. Then in the Renaissance people began to calculate with them, fictitious though they were. In 1673 the great cryptographer John Wallis said they might be imaginary, but were no more impossible than the negative numbers; and now they sort with the reals in the street, drawing never a sidelong glance, although they still bear the caste-mark of their name.

What characterises the living activity of making mathematics is that for something to be a number it must socialize with the numbers already there, able at least to exchange civilities with the natives. It must combine with them in all the familiar ways. For zero to be a power of equal status with what it empowered, we must understand how to add, subtract, multiply and divide with it, for a start: and this was just what the Indian mathematicians did. By doing so, they helped bring about a momentous shift - I won't say in world-views, which have gone out of fashion, but - in paradigms, replacing recipes for a jumble of ingredients with rules for a few simples. As the art of calculation developed an ancestry of theory, zero and numbers evolved toward one another.

Such shifts take place slowly, and are often under way still cloaked in the usages they will outmode. So Brahmagupta in 600 A. D. can on the one hand say quite concisely that any number minus itself is zero; on the other, he struggles toward generality when it comes to adding zero to a number: "The sum of cipher and negative is negative; of affirmative and nought is positive; of two ciphers is cipher" (this translation, from 1817, preserves some of Brahmagupta's lack of distance in its own variety of words for the same thing). He treads with equal and more justified care in spelling out the rules for subtraction:

negative taken from cipher becomes positive;
and affirmative, becomes negative; negative, less
cipher, is negative; positive is positive; cipher, nought.

It would be as if a musician had to learn that G was the dominant in C major, C in F major, A in D major and so on, without ever noticing that the dominant is always the fifth degree of the scale.

Mathematics always follows where elegance leads. Five centuries later Bhaskara rephrased Brahmagupta with elegant economy: "In the addition of cipher, or subtraction of it, the quantity, positive or negative, remains the same. But subtracted from cipher, it is reversed." He wrote this when he was 36, along with a book he called *Lilavati*, "Charming Girl" - perhaps because it was full of problems such as this:

> Beautiful and dear delightful girl, whose eyes
> are like a faun's! If you are skilled in multi-
> plication, tell me, what is 135 times 12?

They don't write math books like that any more.

Mahavira flourished almost halfway between these two, around 830, and thought of zero, you recall, as taking on the protective coloring of what it encountered (doesn't this accord well with the Jain Syadvada, where appearances have no distinct reality of their own?) He also goes on to say that "a number multiplied by zero is zero, and that number remains unchanged when it is... diminished by zero." Brahmagupta before and Bhaskara after agreed.

But my three dots of omission just now trip lightly past an issue on which they seriously disagreed: division by zero. How certain are we ourselves of what the answer should be, and why? Here is what Mahavira says: "A number remains unchanged when it is divided by zero." His translator tries to excuse this false claim by saying that Mahavira "obviously thinks that a division by zero is no division at all." I wonder. Since multiplication could be looked on as streamlined addition (5 x 4, you could say, is 5 4s added together), might he have thought of division as streamlined subtraction (20 ÷ 5 amounts to 5 being taken away from 20 4 times)? If so, when you divide by zero it would be like taking 0 away from that number, which actually leaves it intact. Of course this analogy should have led him to ask how many times 0 can be taken away from 20, not what the result of taking 0 away would be; but as you saw, the backwards thinking involved in subtraction clouds the mind.

Brahmagupta is typically cautious:

> Positive or negative, divided by cipher, is a
> fraction with cipher for denominator [he
> calls this 'khacheda', from that word for
> zero, 'kha']. Cipher divided by negative or
> affirmative is either cipher or is expressed
> by a fraction with cipher as numerator and
> the finite quantity as denominator... cipher
> divided by cipher is nought...

He is certainly right that $0/a = \infty$, where a is a positive or negative number; and saying that a divided by 0 is $a/0$ is only to turn one kind of notation into another without committing yourself to the outcome. But $0/0 = 0$ is just out and out wrong.

Now look at Bhaskara. He begins by keeping his cards as close to his chest as Brahmagupta did: "A quantity, divided by cipher, becomes a fraction the denominator of which is cipher." But then he continues:

> This fraction is termed an infinite quantity
> ['khahara', synonymous with Brahmagupta's
> 'khacheda']. In this quantity consisting of
> that which has cipher for its divisor, there
> is no alteration, though many may be inserted

or extracted; as no change takes place in the
infinite and immutable God when worlds are
created or destroyed, though numerous
orders of beings are absorbed or put forth.

This important passage - which describes a/0 in terms
reminiscent of brahman - has drawn much attention from
commentators. One of them, late in the 16th century, tried
illustrating Bhaskara's meaning by calling up the image of a
sundial. The shadow of the gnomon on it at sunrise and
sunset is infinitely long, he says, and will be so no matter
what the radius of the sundial's face or the height of its gnomon is.

You will still not infrequently hear people say that a/0 =
∞. But is this so? What is that equality supposed to mean?
20/5 = 4 makes sense because it is an equation between
numbers, as Lilavati would have told you; but infinity isn't
a number (it isn't even a stupid number, as children think
who mistranslate the Latin "Infinitus est numerus stulto-
rum": infinite is the number of fools). So what are we supposed to make of a/0? The answer will tell you a lot about
the craft of mathematics.

It would be unimaginably perverse to believe that all
numbers were the same. Experience tells us that 6 isn't 17,
for example (and experience or no, our minds just seem to
come with such distinctions built in). But if you really could
divide by zero, then all numbers *would* be the same. Why?
Our Indian mathematicians help us here: any number times
zero is zero - so that 6 . 0 = 0 and
17·0 = 0. Hence 6· 0 = 17·0. If you could divide by 0,
you'd get 6·0/0 = 17· 0/0, the zeroes would cancel out and 6
would equal 17. They aren't equal, so you can't legitimately
divide by 0. a/0 doesn't mean anything.

This sort of proof by contradiction was known since
ancient Greece. Why hadn't anyone in India hit on it at this

moment, when it was needed? Surely, in part, because proofs, like works of art, come not at our bidding but through the still unfathomed workings of insight; partly because the style of Indian mathematicians was to assert principles rather than prove them; and partly, it may be, because to say something is meaningless is almost like saying you don't know what it means - and as that Arabic traveler Albiruni, whom we ran into before, remarked of the Indians he had met:

> ...they hate to avow their ignorance by a frank
> "I do not know" - a phrase which is difficult
> to them in any connection whatsoever.

On the other hand, the Hindus spoke of Albiruni as "so acid that vinegar in comparison is sweet."

Not only do the Hindus go on to other operations involving zero (Bhaskara correctly asserted that $0^2 = 0$ and $\sqrt{0} = 0$) but they extended the franchise to irrational numbers, such as $\sqrt{3}$, simply by declaring that these could be reckoned with like integers. And as you saw in the Bakhshali manuscript, they legislated for the play of unknowns with numbers of any sort, so that Bhaskara was able to ask: "Tell me, learned sir, the product of 'as much as' five, less absolute one, by 'as much as' three joined with absolute two" - or as we would say, what is $(5x - 1)(3x + 2)$?

While they shied away from imaginary numbers (and even from negative roots of positive numbers - "which people," says Bhaskara in a democratic mood, "do not approve of"), Mahavira was able to form equations like

$$\frac{x}{4} + 2\sqrt{x + 15} = x.$$

Even more impressive, he was able to solve them: for here too, calling such spirits from the vasty deep is one thing; having them come at your call quite another.

What you are witnessing is a language for arithmetic and algebra in the making, whose growth was to have far-reaching consequences. The uncomfortable gap between numbers, which stood for things, and zero, which didn't, would narrow as the focus shifted from what they were to how they behaved. Such behavior took place in equations - and the solution of an equation, the number which made it balance, was as likely to be zero as anything else. Since the values x concealed were all of a kind, this meant that the gap between zero and other numbers narrowed even more. But the real paradigm shift over the long centuries from 500 to 1500 was this: the invisible house of memory, where mathematics had lodged for so long, was giving way to an even more abstract structure. For a multiplicity of names had served to fix numbers and their relations in memorable verse. Now the names would contract to symbols that had to be written down. This made them at once more concrete and even less accessible to non-initiates, because they so abbreviated what they stood for, and also allowed you to say what before you couldn't even think. $x^2 + 3x - 22 = 0$ puts areas (x^2), lengths ($3x$) and constants (22) together in one sentence: hard enough to visualise. But now you could as easily write $x^4 + 3x - 22 = 0$, and solve it - yet how picture the dimension called up by x^4? No wonder William of Malmsbury spoke of "dangerous Saracen magic."

The arcana of mathematical signs lent a glamor to the calling and reinforced the authority and sacerdotal quality it already had. But something much more important was happening within this temple. Like zero, numbers were becoming invisible: no longer descriptive of objects but objects - rarified objects - themselves. "Three" was once like "small": it could modify shoes and ships or sealing wax. Now it had detached itself so far from the rabble of things that instead those ephemera participated briefly in its permanance. Numbers acquired adjectives of their own: positive, negative, natural (the official word for whole or counting numbers), rational (from 'ratio', since these were the fractions), real (rational and irrational) - and in time these adjectives too would become nouns (the Reals, the Rationals). Numbers moved and had their being amid the operations on them, to which even this developing language seemed exterior. All that we saw, all that we sensed, was passing from the causes of numbers to their effects. They were the place-holders, the receptacles, for our counters pushed hectically here and there. By absorbing ourselves in them - by grasping the equation that governs the fall of a sparrow - we could at last bring all the accidents of living into the theatre of thought.

I described in Chapter Four how recursive abstracting is the very stuff of mathematics: no sooner do you weave events together into a coherent network than you reduce that one to a node for another network on a more general level. Let's apply that process here. This enormous paradigm shift in mathematics unfolded, I suggest, within a much broader change of paradigms that took correspondingly longer - beginning perhaps around the fifth century B. C. and becoming firmly established some thousand years later. It was no more than a change in emphasis, as all the most powerful shifts are: a matter of where the accent fell in our analogies.

For we live by likenings. So many of our words are husks of metaphors, so many of our arguments and beliefs proceed by similarities: A is like B. Before this shift began our likenings served, I claim, to enhance and illuminate an A which was too shadowy. If we said that this resembled something else, 'this' was what needed sharpening, while the something else, B, ranged from the differently vivid to the more familiar or intelligible:

> Just as a lion hesitates in a crowd of men, afraid,
> whenever they make a treacherous circle around,
> so Penelope cast her mind about...

That is from Homer, and this from Virgil:

> The dead came thronging round,
> thick as the leaves that fall in the first frost of autumn.

The likenings could be brief as a word or spun out to parables, fables and allegories.

Slowly, however, the accent began to progress, until 'A is like B' was meant to give us some insight into the part that now mattered, B, which wasn't here at all but in an unreachable there. What we saw around us was an imitation or intimation of the Real that lay beyond, below, behind.

> What here we are a god can there complete,

in the words of the German Romantic poet, Hölderlin: and indeed these shifted likenings are the essence of a romanticism, which by this argument, first flowered not in the 19th century but two millennia before.

It takes little effort to unite what seem very different religions and philosophies under the banner that reads: "Not here but there". You find it as early as Heraclitus: "The god, whose oracle is in Delphi, neither affirms nor denies, but points". It is the essence of Plato's vision that appearances fleetingly participate in the Ideas, which are eternal. It is embodied in the Buddhist theory of sunyata, from about the first century A. D., which holds that all entities are empty (sunya) of own being, as they put it ("A new threshold in the history of Indian thought", as one scholar observes). Most familiarly for us, it is the revelation embraced by Christianity and Mohammedanism but resisted by Judaism, the more ancient religion.

When someone asks what this or that means and is given its dictionary definition, you catch the accent shifting in the reply, "Yes, but what does it *mean?*" It is only in recent times - in critiques as various as Nietzsche's and Wittgenstein's - that the image of a world more significant than ours has badly wavered, and the suspicion grown that ours is all there *is* (or, as the accent recedes, that ours is *all* there is).

The change in mathematics we've been following, where names for numbers narrow down to signs of them and the numbers themselves are subordinated to the laws they obey, began when someone first counted and evolved through the on-going project of deriving these laws from as thrifty a set of axioms as mathematicians could manage. In keeping with the shift of the background paradigm, the interplay of numbers came to be understood as manifesting those axioms, which from afar hold taut the fabric of our understanding, like a trampoline's frame. The deep mystery here is that we invent these axioms - their makers have names, the bickerings over them have left scars - yet our conviction grows that we invent them in the only way we can, so that in the end they seem discovered. We experience them as if they lay behind experience; and this both follows from and reinforces the shift in the Great Paradigm.

Of course this shift hasn't rolled relentlessly onward but pools and eddies like a river, and like a river both makes and accomodates itself to its terrain. There are forerunners and backsliders: Plato pointed upward but Aristotle to the variety around him. While Hegel watched Spirit lifting away from Substance, the Panthiests were seeing God in the old oak and the rock. In our own time, as the principles behind things have steadily rarified, Hell - which was intangibly far below everything - has reified (as we know from the nightly news) into the folks next door. The Adam Impulse is alive within us still, making us hope, as the child Darwin did, that plants will tell us their names, and their names will tell us their nature. Yet there has been a shift in the way we have worked our likenings and they have worked on us. For all the fluctuations on its surface, our thinking's center of gravity moved definitively, during the time we've been looking at, from coordinating the meaning of facts to subordinating facts to their significance.

I cannot pretend to call my claim about the Great Paradigm an hypothesis, because it cannot be falsified: any instance either exemplifies the shift or belongs to the retrograde drag on it. Take it then as another likening of the sort I say we have shifted toward. Why so take it? Because it raises the hope (which this shift inspires) of making sense of things.

8

A MAYAN INTERLUDE: THE DARK SIDE OF COUNTING

The Mayan symbol for zero was a tattooed man in a necklace with his head thrown back. Or at least this was one of their astonishing array of zero-symbols. Since their culture flourished from about 300 B. C. to 900 A. D. in the Yucatan Peninsula, isolated from any

0 kins

overseas contacts, they provide vivid testimony to an independent origin for the concept of zero and its signs. And why not? Why should this idea not have sprung up in different cultures, preserved and passed on here, flaring up and dying out there in the minds of many a mute, inglorious Newton? Since mathematics *is* our universal language, obscured by overlays of the accidental languages colonizing the time and place we were born to, why should it not break through with a glory surrounding it, the expression of its constructions altered by those accidents, but the relations so expressed reaching beyond our mortal natures? If we had to give the pedigree of zero as we do that of a horse, we could say: by Imagination, out of Necessity.

Although our image of the Maya is one part fact and three parts conjecture (and these tempered by the fashions ascendant at their making), it is hard not to believe that they counted as if their lives depended on it; and what they counted was time. Their starting-date for the universe would in our calendar be August 13, 3114 B. C. This was their zero day. You wonder how they arrived at it; and since your wonder is unlikely ever to be satisfied, it might turn instead to James Ussher, Archbishop of Armagh, who in the mid-1600s discovered that the world had been created on October 22nd, 4004

B. C., at 6 o'clock in the evening. What a feat! For if you can think at all of the universe beginning, you have to be able to think both sides of that moment (now you don't see it, now you do), a task evidently within Ussher's competence. I like to picture him at his candlelit desk, the shadows of Latin and Hebrew and Greek folios falling about him; bent in perplexity over his calculations; reckoning a final sum; looking up at the flame as revelation dawns - of course! October 22nd! And 6 p.m.! He was much honored in his time and after. For me his dedicated efforts stand as the touchstone of eccentricity.

The Maya scrupulously recorded the dates of important events in terms of their zero day, in what archeologists have come to call the Long Count. Instead of numbering days consecutively from the beginning - a poor use of the brain so avid for patterns - they divided time into years of eighteen months *(uinal)* with twenty days *(kin)* in each. That made a 360-day year called a *tun,* and these tun they then grouped by twenties *(katun),* so that one katun had 20 x 360 = 7200 days. Twenty katun in turn made a *baktun,* or 400 tun (144,000 days); and they had units larger than these, up to an *alautun* of 64,000,000 years. There were distinct glyphs for each of these groupings. In order to show on a monument that the day of its erection was, for example, precisely 1,101,611 days since time began, they would write: 7 baktun 13 katun 0 tun 0 uinal 11 kin (7 x 400 + 13 x 20 = 3060 years, each with 360 days = 1,101,600 days; and 11 days more makes 1,101,611 days). The 7, 13 and 11, like all of their numbers, were made up of bars for fives and dots for ones (so 7 was $\overset{\cdots}{\rule{1cm}{0.4pt}}$, 13 was $\overset{\cdots}{\overline{\overline{\rule{1cm}{0pt}}}}$ and 11 was $\overset{\cdot}{\overline{\overline{\rule{1cm}{0pt}}}}$); but the glyphs for zero - crucial in keeping track of missing middle groupings - were sometimes faces, sometimes full figures, sometimes like half a flower; and in manuscripts at times like snail-shells and at times like nothing we can name.

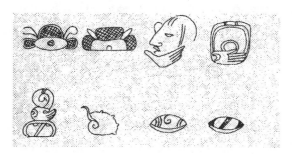

Various Maya symbols for zero

If there were none of the last units, days, the Maya carefully noted this too with a zero glyph (recall that the Greek astronomers didn't): which tells you something was going on more significant to them than getting the date right - something with a formal, ritualistic cast. Keep this in mind as we try to puzzle out the picture of their world.

It has certainly struck you that 360 may be mathematically convenient for the length of a year, but it won't keep up with the recalcitrant nature of things. The sidereal year is roughly 365.242198 days long, so that in less than a katun you would be a hundred kin out in your reckoning of day-length, seasons and anniversaries. Perhaps this was why the Maya had another calendar that ran alongside the Long Count: a "civil" year (the *Haab*) also of 18 twenty-day months but with five "phantom" days tucked in at the end, so that it was only about a quarter day short of the solar year (hence people now slightingly call the Haab the "vague year").

Important as zero was in anchoring the Long Count, it took on a new and peculiar significance with the Haab. The first day of each twenty-day month wasn't numbered 1 but 0; the second 1, and so on, with the twentieth day 19 (the five-day month at year's end followed suit). I don't know of another calendar where the days of the month begin with zero. We rarely even count years in this retrospective way: only the sinister example of Pol Pot comes to mind, since he called 1975 the zero year of his reign - the year in which he would purify his enemies away.

Such numbering always suggests a beginning that is no beginning: a false dawn or overture (so Christopher Robin was all of one when he had just begun). It is the sort of counting that lies behind a particularly clever trick of modern magic. Members of the audience each write secret sentences on slips of paper, seal and toss them into a hat. Blindfolded, the clairvoyant gropes for one at random, with effort intuits its content and announces it. "Yes!" says someone in the audience, "that's just what I wrote!" The mind-reader takes off his blindfold, unseals the slip and reads off the message: it is exactly as he had said. Blindfolded again, he chooses a second, hold it unopened yet somehow, with his third cye, perhaps, is able to see its words and declaims them.

Another person, flabbergasted, acknowledges it. On looking, the mind-reader confirms his vision. And so it goes on, through one slip after another, until all know that they are in the presence of the supernatural. Or are they? It is zero whose power they have felt. The first message the mind-reader deciphered was one he and his accomplice in the audience had earlier agreed on. That was the zero message. When he took off his blindfold what he read on the slip he had randomly drawn was the message he would "intuit" next - and so craftily on, declaring the previous message's content as he read the one that now sat in its stead.

In the Haab calendar, it was on the zero day that the god of the previous month put down his burden of time and the god of the present month was seated, to take it up. For us this occurs in an instant, as when old Janus, keeper of doorways, gives over his weeks to Februa of the purifications. For the Maya this transfer must have been more anxious, as Zip acceded to Zotz or Zotz to Zec. The god who oversaw these events was Zero. A full day for the handing over! I wonder what people did as it dragged by? We know that during the five phantom or "useless" days at the end of the year, the men and women neither washed nor combed themselves, nor undertook any work, lest it miscarry.

Now since the Long Count sufficed to pin down any date precisely and the Haab reconciled the cycle of months to the solar year, what need had the Maya of still another calendar? And yet they had one, and it was very important to them: a sacred year, the *Tzolkin,* of 260 days. Its strange construction tells you a great deal about the Mayan obsession with counting; the concurrence of three calendars points to the source of this obsession in dread.

With this third calendar we can't so much speak of months as of two cycles: one of twenty day-names (Imix, Ik, Akbal...) and one of the numbers 1 through 13. You will appreciate why the Maya held their mathematicians in such high esteem when you learn that the numbers matched the first thirteen names, then began again with the fourteenth: so that the day-names 14 through 20 (Ix through Ahau) went hand in hand with the numbers 1 through 7. Then came the first day-name, Imix, around once more, matched now with number 8, and so through the sixth day-name (Cimi) tied to 13. Manik, the seventh day, was now paired with the number 1 - and this staggering off-beat rhythm went on until 13 x 20 = 260 days after it started, when Imix stood for the first time again with the number 1 and a fresh sacred year commenced. Any combination of number and name would tell you precisely which day of this year it stood for - if you had the knack of it through insight or arduous training.

Why such a bizarre calendar? Perhaps because for the Maya there were thirteen gods of the upper world, while twenty was the number of man (not an uncommon assignment, given ten fingers and ten toes). This interlacing of the two cycles may therefore have been meant to harmonize the secular and the divine. There were also nine gods of the underworld in the Mayan pantheon; they had skeletal lower jaws and were ruled by the Death God. So is it any wonder that they had a fourth calendar too, a cycle of nine glyphs representing the Lords of the Night who ruled each day in turn? And a lunar calendar of 29- and 30-day months; and a sixth calendar, based on the 584-day synodic cycle of Venus (its apparent oscillation from one side of the sun to the other)?

Although theirs is the only culture to have institution-alised such a complex obsession with counting, surely each of us has succumbed at one time or another to its compul-sive insistence, telling off the seconds, the cracks in the side-walk, the patternless holes in the dentist's ceiling tiles; and always what begins in jest ends in earnest, needing no little effort to turn off the machine as we begin to count its counting, to doubt it and double back, to lose our way and wish it done with, in vain, until we recognise that servant and master have changed places and we are no more than the housing for a relentless automaton. Some have reached an accomodation with their monster: Sir Francis Galton, cousin of Darwin and the father of Eugenics, counted everything in sight and even had gloves made up for him with pistons that drove ten separate counters, so that he could unobtrusively keep track of the percentage of beauti-ful women in Macedonian villages while tallying up the average price of goods in their shop windows. Others have just given themselves up, like the otherwise lumpish farm-hand Jedediah Buxton, who in the eighteenth century couldn't help calculating how many hair-breadths wide was every object in his path; and who, when taken to London as a treat to see the great Garrick in a play, announced at its end precisely how many words each actor had spoken, and how many steps they had taken in their dances. We hear tales linking autism's estrangement to a preternatural fasci-nation with counting; we remember Freud, in his old age, insisting that rhythmic repetition was the expression of our longing, even beyond pleasure, for sameness, and sameness was the emissary of death.

We can't, however, dismiss the Maya as suffering from some sort of collective arithmomania, since it wasn't so much the counts as their coincidings that mattered to them.

And here once again we have to admire their mathematical skills. If the 365-day Haab and the 260-day Tzolkin began together, when would the first day of each next be the same? Put the problem with more manageable numbers to see how to solve it: when will a two-day and a five-day cycle coincide? Clearly after 2 x 5 = 10 days: 10 is their least common multiple. And 4- and 6-day cycles? 4 x 6 = 24 is their product, yet 24 isn't the answer: they will first coincide after 12 days, because you have to divide the product by 2, since 2 is the greatest common divisor of 4 and 6. The Maya understood that since 5 was the greatest common divisor of 260 and 365, the Haab and the Tzolkin would begin again together after

$$\frac{260 \times 365}{5} = 18{,}960 \text{ days,}$$

which was 52 Haab or 73 Tzolkin years. This period is called the Calendar Round, and it seems that each completion brought unmitigated cruelties. Why? Let me add a light patina of unifying conjecture to the scatter of those that shape our sense of Maya thought.

Their great fear was that time might stop; for since it carried all its sons away to the pulse of the Long Count, why should their empire, the earth, the sky with its countless stars and the universe itself not perish as well? To prevent this they took drastic measures, both clever and horrible. First, they transferred the cyclicity they saw in the heavens to linear time, whichg therefore couldn't stop in the middle of a cycle - but might yet at its end. Very well, set in motion another cycle out of phase with the first: now time couldn't stop until those rare occasions when the ends of both cycles coincided - and at such evil moments (as every 52 civil years), they offered up to the gods the essence of vitality: blood, virgins, hearts cut from living victims, that the gods might revive themselves with these and so be willing to pick up again the burden of months put wearily down on zero-days.

The trouble with obsessions is that once in their grip, nothing is ever enough. More cycles, more cycles, to put off the danger to yet ever greater multiples! 5 synodic years of Venus made 8 years of the Haab; 405 lunations were 46 Tzolkin years; their codex now at Dresden lists multiples of 78, and the synodic year of Mars is 780 days; one archeologist even suggests that since 9 was the number of gods of the underworld and 13 of the upper, and since 9 x 13 = 117, the Maya must also have calculated the synodic year of Mercury, which is 116 days (keep in mind how vivid and portentous the night sky must have been to people from whom it wasn't hidden by city lights). On the other hand, there were the divisions of cycles to worry about: every five years (the quarter-points of a katun) their king spectacularly mutilated himself that his blood might keep the thirsty gods at their task. As a pair of scholars put it, blood was the mortar of ancient Maya life.

Nothing is ever enough. Their Long Count held out the promise that time wouldn't stop for an immense span indeed. A system of dating we find on some of their monuments acted like the service coupons that come with your new car: if there is a voucher for a 50,000-mile check-up you feel confident that your car will just be coming to maturity then. If there is none at 150,000 miles, a shiver runs through you. This system of theirs seemed to show that they had hardly moved forward from time's beginning in a stretch (by one reckoning) of about 2×10^{27} years. By comparison, our own culture's Big Bang was a puny 12 or so billion years ago (1.2×10^{10}). But like some post-modernist novel of drawers within drawers of a tumbledown desk, even the Long Count itself became thought of as no longer linear but as the latest period of these immensely long spans, which had reached cyclically back and would reach cyclically forward forever. It was as if the Maya had taken out insurance policies on their insurance policies.

A wonderful advantage lay in rethinking the Long Count as itself periodic. Not only would the threat of time's cessation diminish, but now the birth of a great ruler could be proven (if your mathematicians were agile enough) to have fallen precisely a significant number of years after, say, the birth of a mythical ancestress - the significance being that the time-span was an integral multiple of this cycle and that: the Haab and the Tzolkin, the synodic period perhaps of Mars and of Mercury to boot. Such good fortune in catching the crest of the polyrhythms was apparently taken to show that the birth was a rebirth: the human Pacal, greatest of all Palenque's rulers, *was* his divine ancestress.

Once you let the analogy "A is like B" turn into "A *is* B", imagination hardens into the excesses of fantastic conviction, and the catch on the box that stores the demonic snaps open, letting Hell loose. I mentioned that the gods of the underworld, the nine Lords of the Night, were ruled by the Death God - but I didn't tell you who this death-god was: he was Zero. His was the day of the Haab when time might stop. His was the end of each lesser and greater cycle, fearful pause. Now if a human were found who could take on Zero's persona - and if *he* could be put to ritualistic death - then Death would die! And this, it seems, is just what the Maya did. They had a ritual ballgame between a player dressed as one of their hero twins, and one dressed as the God of Zero. The ball was an important hostage, such as a defeated king, who had been kept for many years and was now trussed up for the occasion. The two players skillfully passed and kicked and beat him to death, or killed him in the end by rolling him down a long flight of stairs; and it was the hero twin who always won by outwitting Zero. In other such games, the loser was sacrificed. But outwitting death wasn't enough. A human would be dressed in the regalia of the God of Zero, and then sacrificed by having his lower jaw torn off. As with most religions, the failure of ritual to achieve its aim didn't alter it, since even the barbarous live in hope.

Anthropologists of another generation, following Ruth Benedict, would have called the high culture of the Maya dionysian - but that, I feel, is a slur on Dionysus. This blood-soaked society, with its glyphic wit, so clever at numbers, so artful in building, so apt at astronomy, reminds me of a brilliant and very neurotic friend I had who appeared one day without his tics and twitches. What had happened? "I traded all my neuroses in," he explained, "for one little psychosis, and now my life makes sense."

9

MUCH ADO

1. ENVOYS OF EMPTINESS

Not all their cruel ingenuity kept Zero, the God of Death, from triumphing over the Maya at last. As their culture decayed, another, half a world away, was spreading. Arab merchants were carrying exotic goods, tales and techniques in every direction. We tend to abbreviate them to their eyes, fixed always on stars or barren horizons. But the emptiness they brought wasn't a whiff of the Empty Quarter (only disappointed romantics meet none but themselves in the wastes: for those who live there, the desert is full of incident and accident, with as many kinds of dune as of camel). Rather, it was the zero of positional notation which they had found in India and which was in Baghdad along with the other Hindu numerals by 773 A. D., and which gave them such fluency in estimating, bargaining and reckoning - for while they may have used these numerals to record results arrived at on the counting board, they were calculating directly with them by around 825, when Al-Khowarizmi wrote his work on arithmetic. Their commerce wasn't only material: with their religious fervor went a youthful admiration for the variety of learning and invention their travels uncovered. These goods of the mind, translated into a common tongue, spread from the new seats of learning in Damascus, Baghdad and, later, Cordoba as broadly and rapidly as silk and steel.

The routes could be circuitous. The Hindu numerals reached southern Russia, for example, by way of Ismaelite missionaries from Egypt, around 990 A. D. - as we gather from Avicenna's autobiography. He was a man who deserved to write one. Born in Bukhara, at ten he learned arithmetic from the greengrocer and by seventeen had read Aristotle's <u>Metaphysics</u> forty times - but first understood it only after picking up a small commentary for a penny. He was the most famous physician of the middle ages, equally renowned for his heroic bouts of sensuality as for his learning.

It could have been Arab merchants on the ivory and spice roads, or even earlier, Buddhist and Hindu travellers, who brought zero with them to China - for in her novel, <u>The Hundred Secret Senses</u>, Amy Tan's holy innocent, Kwan, was wrong:

> "Course, probably Chinese invent pencil, we
> invent so many things - gunpowder but not for
> killing, noodle too - Italian people always say
> they invent noodle - not true, only copy Chinese
> from Marco Polo time. Also, Chinese people in-
> vent zero for number. Before zero, people not
> know anything. Now everybody have zero."

The Indian ancestry of the Chinese zero is shown not only by its forms - dots, perfect circles, even circles within circles - but by 'ling', a character for zero, which meant the last few raindrops after a storm, or those that cling like beads to leaves and umbrellas: for remember one of Mahavira's many synonyms for zero, 'nabhas', vapor; and even more significantly, 'bindu' itself, which also meant droplet or globule.

There was a time, Kwan, when scholars toyed with the idea that zero arose from the emptiness of Taoism, which it then mingled with the Hindu sunya (if you can picture mixing one hollow with another) - but that time is past, and the phrases which lay like parsley around the conjectures - "It may be...", "We are free to consider the possibility..." have interlaced and smothered them away.

I'd like to think (moving my own parsley judiciously to the rim of my plate) that the board used in what is arguably the most popular game in the world, Mancala (or Kalaha, or Warri, or any of the hundreds of names from Abalala'e to Yovodji by which it is variously called) first came to Africa in the saddle-bags of Arab or Indian merchants, as their counting boards. For if these had not ruled lines but the rows of pits you see in this game, it would buttress my conjecture that zero's hollow circle had its origin in such shallow depressions: now not the impress left by a stone removed from sand, but the places themselves that counters

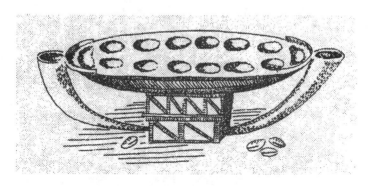

A Mancala Board

lay in or moved through (so that an empty pit would be the aptest symbol for zero). It certainly isn't exceptional for a game to have had such a past: think of checkers and the exchequer's counting boards; for when serious usages are surpassed - as was the board by the 'Arabic' numerals - they enter a second childhood and become playful again. The antiquity of these Mancala boards might even make you wonder whether they might not have been coeval with, or pre-ceded, the ruled board. Such boards, they say, are carved at the base of columns in the Temple of Amon, at Karnak, and can be seen as well in the rock ledges along ancient caravan routes. I will resist the temptation to compare the cowrie shell commonly used with them to the all but identical form of Mayan zero long thought to depict some sort of shell: for that way mad speculation lies.

Zero came West certainly by 970, perhaps as early as a century before, resplendent with names from a mixture of sources, some drawn from its sense, some from its shape.

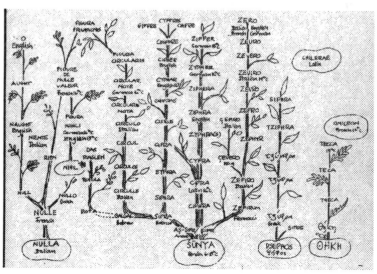

Some names for zero

Most displayed their lineage in their etymology, a few wore deceptive plumage - but all at last fluttered into the branches that fork from the Middle Ages to our day. They were birds of a feather whose iridescence made them seem an aviary.

The ancestor of so many of these Western names was the Arabic sifr or as-sifr, itself a translation of the Indian sunya, "void"; but the Greek psephos, "pebble", for "counter", added its weight here and there, and their theca, "receptacle", spawned descendants of its own. Those wonderful coincidences of sound and sense between one language and another was also at work, giving each new term an alluring resonance. So the Hebrew "sifra" was allied to sifr while having its own connections, perhaps, to words for crown and counting. Various medieval Latin names for circle on the one hand (such as rotula and circulus) and emptiness on the other (nulla, nihil) had their begots. The languid zephyrs that turned and returned over Italy in its spring, *zefiro, zefro, zevero,* weakened by the time they reached Venice to the *zero* that is now ours.

2. A SYPHER IN AUGRIM

Were these the names for no number at all or for numbers in general, as "chiffre" in French and "Ziffer" in German became? Did they stand for the merchant's public art of counting, as in the English "ciphering", or for the secret writing of spies (an undertone preserved for us still in the words "encipher" and "decipher")? Certainly the confusion stood for trouble ahead. I imagine there were three sources from which this trouble flowed, their waters braiding and broadening: superstition, bafflement and distrust.

Anything imported into what was still largely a peasant culture in the West would likely have been looked at askance; anything from the East was especially dangerous,

seat as it was of old and still potent heresies. Most hated
and feared of these was Manichaeism, that third century A.
D. mixture of Persian mythology and Gnostic theology,
which lasted in various forms through the middle ages. It
saw good and evil in equal struggle, God and the devil
fighting it out on the battlefield of man. As conclusions
accumulated to a system, two features stayed constant that
matter for us: the first, that the void was identified with
evil; the second, that forces and beings could be *evoked*
into existence: they could be called forth by naming. When
the Manichaean dualism was chased into hiding by the pas-
sions which conflicting beliefs let loose, what emerged from
darkened corners were odd observances, rituals of avoidance and invocations to shadowy figures in times of need.
Substance had turned to superstition, all the more powerful
for being unexamined.

To the extent, therefore, that zero was connected in
shape or meaning with the void, it had to be dealt with gin-
gerly, if at all. A Roman idiosyncrasy about counting reinforced this avoidance. The 360 degrees of longitude, for
example, were always measured from the vernal equinox,
which lies in the zodiacal sign of Aries. This should be zero
degrees, 0°. It was common, however, to call it instead "the
first degree", Aries 1°, as Pliny did around 60 A. D., upset-
ting his calculations and those of many who followed him.
It amounts to this: if you lay out four marks on the ground
and step from the first to the last, have you taken three
paces or four? Clearly three; yet four marks were involved.
To get the right answer it helps to call the starting-line
"zero"; then the number at the mark you step on will cor-
respond to the number of your steps. But the Romans
counted so that three days after Sunday was Tuesday; the
Italian for the 15th century is the quattrocento - and all of
us still call the two steps in music from C to E a major
third, taking the number from the three tones involved.

Superstition made zero abhorrent to the godly, while bringing it into the arcana of those who crossed over to the occult. In alchemy its shape appears as Ouroboros, the dragon that swallows its own tail, symbolising the *prima materia;* and as rota, the circular course of the alchemical transformation. Its circle is everywhere in magic, marking off spellbound earth. Nor is this circle confined to rural rites and party conjurors: it appears again as the mandala in the magic-ridden psychiatry of C. G. Jung. In fact whenever we grow dissatisfied with our wits and think them no match for darkly glamorised forces; whenever ancient wisdom glows with a fiercer light than modern knowledge; whenever the petty distinctions among things diminish and we realise that anything can be everything and each is also its opposite - then the image of zero's perfect ring shines before us: zero, the number of The Wise Fool in the Tarot deck, and the anagrammatic Tarot itself, transformed to Rotat.

The Arabic numerals collectively provided initiates with one more set of symbols to bind them exclusively together, and to manipulate in ways that might summon up who could say what powers. You can recapture a sense of the perplexity, tinged with awe or dread, that a layman must have felt on seeing a run of astrological writing:

What instructions were these, what predictions? They are just the numbers one through six from *The Book of Numbers* written by Noviomagus in 1539. He claimed

Noviomagus' first six numerals

them to be Chaldean, but they bear no relation to any numerical system we know.

Even for those immune to superstition, zero was a number "donnant ombre et encombre", as a fifteenth-century French writer put it: a shadowy, obstructive number. What it was, how it acted - but above all, what it *meant* - was baffling. For anything with a name (and zero had so many) surely existed: not only the Manichees believed that names denoted real things. Yet how could what didn't exist, exist? To the objection that God could not have created the world out of nothing because the infinite distance between nothing and being couldn't be crossed, Thomas Aquinas weakly replied that thinking of creation as a change between two terms led to falsely imagining such a gap.

Even if you tried to ignore what zero meant, you couldn't ignore how it and the place-holding system were supposed to work. This is so easy for us, the inheritors of half a millennium's practice; but it was mountainous to many a scholar in the middle ages. Picture yourself trying to learn it from the earliest English text on the subject we know, *The Crafte of Numbrynge*, written about 1300 (I have modernised the spelling):

> This book is called the book of algorim or augrym after lewder use. And this book treats of the Craft of Numbering, the which craft is called also Algorym. There was a king of Inde the which was called Algor and he made this
>
> craft.... Algorisms, in which we use the figures of Inde... Every of these figures betokens him self and no more, if he stand in the first place of the rule...
>
> If it stand in the second place of the rule, he betokens ten times him self, as this figure 2 here 20 tokens ten times

him self, that is twenty, for he him self betokens twain,
and ten times twain is twenty. And for he stands on the left
side and in the second place, he betokens ten times him self.
And so go forth...

A cifre tokens nought, but he makes the figure to betoken
that comes after him more than he should and he were away,
as thus 10. here the figure of one tokens ten, and if the cifre
were away and no figure before him he should token but one,
for that he should stand in the first place...

Perhaps you now feel less confident than you did about King Algor's system - especially when you find there was no such king, but 'augrim' and its like derived from Al-Khowarizmi's name (the source of our own 'algorithm'). But if you think the efforts of our anonymous writer laughable, put yourself in his place by trying to explain a recent notation devised by the mathematician Donald Knuth. It is needed for dealing with the extraordinarily large numbers that come up in Ramsey Theory (a branch of mathematics that studies rapidly growing functions). It begins easily enough with $3\uparrow 3$ meaning just 3^3, or 27. $3\uparrow\uparrow 3$ is to be understood as $3\uparrow(3\uparrow 3)$, that is, 3^{3^3}, or 3^{27}, which is the sizable number 7,625,597,484,987. So $3\uparrow\uparrow\uparrow 3$ is $3\uparrow\uparrow(3\uparrow\uparrow 3)$, that is, $3\uparrow\uparrow 7,625,597,484,987$, i.e., $3\uparrow(7,625,597,484,987\uparrow 7,625,597,484,987)$.
And $3\uparrow\uparrow\uparrow\uparrow 3$ hym selfe betokens $3\uparrow\uparrow\uparrow(3\uparrow\uparrow\uparrow 3)$ and for this figure 3 standis on the lyft side

of ↑ 3. &... But perhaps we should stop here, before past and future meet at infinity.

Say, however, that you grasped how to use positional notation. Now something even more baffling blocked your path. These were not the times of Solon or Polybius, when status was as much at a ruler's whim as the value of counters on a counting board was at the casual hand of the reckoner. Order was hierarchical, from the village to the universe; hierarchy gave the world its sense and you your place within it. Yet equal in importance to hierarchy was allegory. In this ultimate epoch of the Great Paradigm, everything pointed past itself to its significance. What then could this fluidity of position, brought from the East, point to? With hindsight we can confidently answer: it pointed to change; to the end of the long stasis. Just as pictorial space, which had been ordered hierarchically (size of figure corresponded to importance) was soon to be put in perspective through the device of a vanishing-point, a visual zero; so the zero of positional notation was the harbinger of a reordering of social and political space.

That wasn't how it was read at the time. The best minds were perplexed. Even as late as the 1620s, John Donne could say from his pulpit: "The less anything is, the less we know it: how invisible, how unintelligible a thing, then, is this *Nothing!*" Some two hundred years earlier another Englishman, Thomas Usk, had come to a slightly more positive conclusion in his *Testament of Love*: "Although a sypher in augrim have no might in signification of itself, yet he giveth power in signification to other."

A wonderful understanding of this positional zero, wholly within the terms of the medieval canon, did emerge and flicker briefly in Cologne, early in the fourteenth century. There Meister Eckhart, the father of German Idealism - a Dominican, a radical mystic, a man who preached in the vernacular with confounding eloquence - taught that all creatures are nothing; that being empty of things is to be full of God; that God, who must lie past all knowledge and all Being, must therefore also be nothing, has been immovably disinterested in his creation from the beginning, and still is - and disinterest *(Abgescheidenheit)* comes so close to zero *(Nihte)* that nothing but God is rarified enough to go into it. Is this a message of Stoical resignation in a god-forsaken world? Far from it: Eckhart has a very different vision in mind, a vision in which he sees that he is God, and that anyone will be God if he goes beyond humility to disinterest. In the course of delivering one of his last sermons he announces the truth which that moment was discovered to him: "God and I are One. Now I am what I was and I neither add to nor subtract from anything, for I am the unmoved Mover, that moves all things." And that, of course, is just what zero is.

Even if you accepted the Arabic numerals with their zero for what they were and what they signified, how could you entrust your calculations to them when, compared to the counting board, these were so hard to do? Addition was relatively easy, once you came to grips with carrying; subtraction put your abstract powers to the test. Were you going to learn to multiply 'by the cross', 'by the fold', 'in columns' or on the diagonal, 'in the manner of a jalousie'? Whichever you chose it would take much skill ("a lot of brains", as the German reckoning-master, Adam Riese, said in the 1500s).

If you were a merchant in Germany and took this matter seriously, you sent your son to Italy where they ordered these things better (a Nuremberg father wrote to his son in Venice that he hoped he would learn to rise early, would go to church regularly and would master arithmetic). Dividing was difficult on the board - so difficult, in fact, that one way of doing it was called "iron division" (divisio ferrea), because it was "so extraordinarily difficult that its hardness surpassed that of iron." But with Arabic numerals it was just as confusing when you divided by the strike-out method: this left your sheet covered with crossed-out numbers in rows that grew and shrank, the outcome looking just like a ship under full sail, leading the Italians to call this technique "divisione per galea", 'division by the galley'.

Mistakes were so common that in French "faire par algorisme", "to do it with algorism", came to mean "to miscalculate": a sensible conclusion, when pencils were scarce and paper scarcer, so that you crammed what calculations you could into stray corners that left no room for accuracy. How were you supposed to fly or even crawl through a problem such as this, from a textbook of 1489:

A man goes to a money-changer in Vienna with 30 pennies

in Nuremberg currency. So he says to the money-changer,

"Please change my 30 pennies and give me Vienna pounds

for them as much as they are worth." And the money-changer

does not know how much he should give the man in Viennese

currency. Thus he goes to the money-office, and they there

advise the money-changer and say to him, "7 Vienna are worth

9 Linz, and 8 Linz are worth 11 Passau, and 12 Passau are

worth 13 Vilshofen and 15 Vilshofen are worth 10 Regensburg

and 8 Regensburg are worth 18 Neumarkt and 5 Neumarkt are

worth 4 Nuremberg pennies." How many Viennese pennies do 30

Nuremberg pennies come to?

You might care to know that the answer is 13 23/429 (round it off to 13 Viennese pennies, with the rest left as a tip for the advice).

Some teachers would have none of it. Prosdocimo de' Beldomandi, writing around 1400, said that he found many different techniques in many different books, but in all of them if you made a mistake early on you would have to begin all over again; and where could you store up partial results, and how could you erase them? It was all too laborious and fastidious for him, and so he would throw it all away and include in his book only the bare minimum necessary for calculating.

These technical difficulties, combined with the slow spread of knowledge before books were printed and writing in the vernacular was common, added to the reputation that the Arabic numerals already had for being dangerous Saracen magic. Even when they began to appear as dates on coins and monuments, banks were still reluctant to use them, and for good reason: zero was the villain again, since it could be turned into a 6 or a 9 by the unscrupulous, who could also slip in a digit or two before it. So in Florence the City Council passed an ordinance in 1299 making it illegal to use numbers when entering amounts of money in account books: sums had to be written out in words. An old Venetian text on book-keeping explains that "the old figures [i.e., Roman numerals] alone are used because they cannot be falsified as easily as those of the new art of computation." At the University of Padua the stationers were required to write the prices of books "non per cifras sed par literas claros": not in figures but in clear letters. In 1494 the Mayor of Frankfurt instructed his master calculators "to abstain from calculating with digits." Even as late as 1594, a canon in Antwerp warned merchants not to use numerals in contracts or drafts. We laugh at those who can't count - but in the 13th century they laughed at those who could, making 'cipher' and 'the zero of algorismus' terms of derision, because of their uselessness:

A horned beast, a sheep,
An algorismus-cipher,
Is a priest, who on such a feast-day
Does not celebrate the holy Mother.

3. THIS YEAR, NEXT YEAR, SOMETIME, NEVER

This unruly crowd of square-headed suspicions jostled and shouldered zero aside as it struggled to make its way westward. A subtler and more powerful combination of concerns had long since been aligned against it, for as the centuries heaped up it became increasingly important to know when the world would end.

Roman-style counting confused the issue, since there was no year zero between 1 B. C. and 1 A. D.; hence millennialists had to reckon then - as they do now - with the difficulty that years ending in zero were the last of their decade, century or millennium, not the first of the next (so for us, January 1st of 2001 begins the third millennium, and the festivities of the year before celebrate only private rites of passage). Despite numerous calendrical reforms, it wasn't until 1740 that a Year 0 hesitantly made its appearance, when Jacques Cassini, the second of four generations of Italian astronomers in charge of the observatory at Paris, published his *Tables astro-nomiques* with this rectification (he too must have had a sense of zero as the unmoved mover: like his son and grandson, he was born in the observatory, from which for over a century the family surveyed the skies and the mellow fields of France).

To know when the end would be: for this you needed to date exactly the relevant beginning, and understand what span of after-time was ordained. This is why such efforts of exegesis and calculation went into establishing Christian chronology, whose purpose - as that astute scholar of things millennial, Richard Landes, observes - "had ever been to date not the beginning but the End." We have seen these calculations, for all the precision with which they were made, miscarry time after time: the Dukhobors burned down their houses, with all their worldly goods, the night before recurrent Judgment Days; people stood with their prophets at dawn in mountain meadows, only to trudge disappointed home again at dusk. Hale-Bopp came - and went.

Yet these crushing disappointments left behind residues of incalculable value: an altered sense of time and a wakened eye for history.

Dies irae, dies illa
Solvet saeclum in favilla -

Day of wrath, that day
Will crush the years to cinders -

If the zero day had safely passed, then time before and time after took on new meaning: each extended away as the negatives and positives do from zero, now no longer the terminus but the fulcrum of counting. The past needed closer study for signs of what was to come; and these broken apocalyptic promises led, says Landes, to a discourse that sought "to restructure a future with longer temporal horizons": the more distant The Day, the more it could be dwelt on in hope. Of course it shouldn't be so distant as to discourage nor so near as to disappoint, but should lie in a temperate zone of time, at least a generation away but no more than three or four.

Here too, however, the Great Paradigm was at work. The actual end of time: the day when, as St. John foretold, a small voice would be heard saying: "A measure of wheat for a penny, and three measures of barley for a penny", and the heavens would roll up as a shut scroll and the stars would fall like unripe figs and the four horsemen would be loose in the land. An actual day in 1000 A. D., or in 1260, 1533 or 1843: to believe this is to let loose the very real demons within and around us, burning and looting, because there would be no tomorrow. But calmer heads knew that when tomorrow came, the price to pay would be gigantic.

Not just to avoid this tearing of the social fabric but to redirect thought from the word to what it means, from matter to what matters, authoritative voices within the church had long since spoken out against predicting when the apocalypse was to be. As early as Matthew and Mark we are told that not even the angels know this date. St. Augustine, and after him a whole tradition of writers, including Bede, worked in a variety of ways to put the dreaded day off or aside. Different systems of reckoning, from different datum-lines, were conjured up; errors were found in calculations that brought the moment too near. Cleverest of all, as with that fascinating monk, forger and historian, Ademar of Chabannes (who wrote in the millennial generation between 1000 and 1033), chronology was deliberately allowed to become vague around pointers to the critical time: Ademar substituted expressions such as "in those days' and "after a while" for the precise dates in the annals he was copying.

An upshot of these warring traditions - the prophets of doom stronger in speaking, their rivals in writing - is to drift yet one more wisp of fog over the already obscured nature of dating, with its initial and terminal zero missing or shifted or blurred, and so driven underground. When it came knocking again at the door, little wonder that those within were uneager to answer.

You can't help hearing this story with the Maya in mind - yet how differently the taut string of time resounds in the two. The Maya, as you saw, feared that time was linear and so could end. To prevent this they imposed on it one exactly delineated cycle after another, thinking thus to force it recurrently on. Christians were certain that time was linear and so would end in a Day of Judgment, bringing them into timelessness with their god.

What then of the cycle of seasons, of days and weeks and months? These were either devalued as pagan relics (the moon, whose phases meant so much in the ancient world, now only sorrowed with the sun at the crucifixion, or supported the Virgin Mary in the sky); or were revalued, as in the bleak midwinter fires of Beltane, co-opted into the celebration of a singular birth at Christmas. Cycles lingered here and there, as in telling your beads; most often, however, they were uncoiled into rhythmed but forward-falling time. So the week imitated the seven days of creation, on whose sabbath your soul and heart and mind were to be bettered through new, not renewed, observance. The structure of Western music well illustrates the subordination of the cyclic to the linear: melody moves, with sabbatical pauses, to an end different from its beginning, through patterned repeats of harmonic form. From capo to coda, head to tail, each piece is a singular creature.

The contrast between the linear symbolism of the dominant faith (the stations of the cross, the soul's journey) and the cyclic symbols belonging to the beliefs it had covered (returns seasonal and eternal) led at times to baroque compromises or conflations: think of Isaac Watts' version of

Psalm 90: "Time, like an ever-*rolling* stream..." - or much earlier, the tondo and halo alongside the cross. The pilgrim to Compostela, after all, carried both a scallop-shell and a staff. But the tension between the two outlooks, never fully resolved, added a last ambiguity to the medieval nature of zero, inheritor now of the different characters it bore in these opposed scenarios.

4. STILL IT MOVES

It is tempting to borrow some of this linear imagery and say that, like Goethe's Faust, zero came through its dark struggles by cleaving to the one true way. But history walks on human feet. Three figures stand out among those bearers who left zero's mystic and misty veils to snag on the riddles of the day and brought it unconcealed toward the Renaissance: a man from the House of God, a man from the Sacred Wood, and the son of the good ol' boy.

Alexander De Villa Dei wrote "The Song of Algorism", *Carmen de Algorismo*, in 1240, and John Sacrobosco his *Algorismus vulgaris* in 1250. Each was enormously popular in the universities: read, copied and commented on, line for line. Lecture notes, taken down in shorthand, still exist from the 13th century, repeating that this science was due to the philosopher Algus, and forming the symbol , teca, for zero. Some commentators confessed that they couldn't quite understand certain passages of unwieldy classification, and surely in the small hours of the night many a university scholar scratched his head over Villa Dei's verses:

Prima significat unum; duo vero secunda;
Tertia significat tria; sic procede sinistre
Donec ad extremam venias, quae cifra vocatur.

Fibonnaci

"The first means one, the second truly two, the third means three, and thus proceed to the left until you reach the last, which is called cifra."

But it was the son of the good ol' boy, Filius Bonacci - Fibonacci for short - whose studies went deepest and had the greatest effect. His proper name was Leonardo of Pisa but he called himself Bigollo, which some have since thought to mean "blockhead", others "the traveller". He was no blockhead, but his face was that of an ironist; and he did travel as a merchant through Egypt, Syria, Greece, Sicily and Provence in the last years of the 12th century, observing, inquiring, comparing and bringing back with him what he saw. In 1202 he published his confusingly-titled *Liber Abaci*, "The Book of the Abacus", which wasn't about the abacus (i.e., counting board) at all but the Arabic numerals, the best of the calculating systems he came on.

Unlike Villa Dei and Sacrobosco, Fibonacci didn't just transcribe the new system of counting but played with these toys as a mathematician would. The sequence ran 0, 1, 2, 3... Very well: he experimented with another that started 0,1, 1, 2, 3, and went on: 5, 8, 13, 21, 34, 55 - because each term was the sum of the previous two. The foolish fancy of a blockhead? This Fibonacci Sequence has since been found everywhere in nature, from the divisions of a nautilus shell to the cross-hatchings on a sunflower head. A historian of mathematics once wrote that not a professor at the University of Paris, then the intellectual center of the world, would have understood the fine reasoning of this little merchant from Pisa.

Yet progress seemed to take one step back for every two forward, as if it had sworn a pilgrim's oath or was playing a children's game. The work of Brahmagupta, Mahavira and Bhaskara to put zero on an equal footing with other numbers didn't survive the voyage westward: Fibonacci speaks of the nine Indian *figures* but the *sign* 0. It isn't until the Crafte of Nombrynge in 1300 that we read:

...ye most undirstonde that in this craft
ben vsid teen figurys, as here bene writen
for ensampul, 0 9 8 ∧ 6 4 ℵ 3 2 1 ... in the
quych we vse teen figurys of Inde...

Almost another two hundred years passed before zero was treated like an actual quantity: in 1484 a physician in Lyons, Nicolas Chuquet, solving for x in the quadratic $3x^2 + 12 = 12x$, found by his method that $x = 2 \pm \sqrt{4 - 4}$, and remarked that "since 4 -4 = 0, $\sqrt{0}$ added to or subtracted from 2 leaves 2, which is therefore the number we sought." But his work, *Le triparty en la science des nombres*, wasn't published until after his death, so that even more time elapsed before zero's green-card was replaced by a naturalisation certificate.

Meanwhile the rising tide of commerce swelled the demand for careful calculations and records of transactions. The Arabic numerals were taught but, as you saw, distrusted. Archaic yet wonderfully clever devices persisted, such as the tally-stick, used even as late as the 18th century in England (stacks of them catching fire a century later burned down the Houses of Parliament): notches cut in its length showed the sum owed; the stick was then split down the middle, one part kept by the debtor, one by the creditor: their unique tallying prevented fraud. Millers showed the kinds and amounts of flour in their sacks by the way they knotted the draw-strings; and everyone, merchant and banker, sophisticates and illiterates, knew how to reckon sums up to a million with their fingers.

But the counting board still was the champion to be defeated. As feudalism fell away and the uncertainties of rank in a more dynamic society returned, so did the old simile of courtier to counter. At first it contrasted the stability of hierarchy Here to equality There:

> To the counting master all counters are
> equal, and their worth depends on where
> he places them. Just so are men equal
> before God, but they are unequal according
> to the station in which God has placed them.

So Martin Luther. Two centuries later and a country away, however, we have:

> Les courtisans sont des jetons,
> leur valeur dépend de leur place:
> Dans la faveur, de millions,
> et des zéros dans la disgrâce!

Which is:

Courtiers are but counters,
Their value depends on their place:
In favor, they're worth millions,
And nothing in disgrace!

(The way this enduring metaphor acts as a barometer of social stability might almost yield an essay, where you could also savor Plutarch's variation: "Just as, in calculating, fingers sometimes have a value of 10,000 and sometimes just 1, the favorites of kings may be either everything or almost nothing." Observe the Roman ignorance of zero. On the other hand, observe our French aphorist's apparent ignorance of how a counting board works).

The decisive blow in this battle fell before anyone realised it: and it was a blow decisive as well for the course of zero and of Western thought. What happened was this. The need for accurate accounting amidst burgeoning data led to the invention in Italy of double-entry bookkeeping some time before 1340. As with all great inventions, the idea was simple: tote up your credits and debits on the same page of your trading account's ledger, in parallel columns. If the difference between them is zero, your books are balanced, showing your accounts were accurately kept (or skillfully cooked). And profits and losses? These too were doubly entered there by creating a second, nominal account of profits and losses to which the profits, say, from the first account were transferred, this transfer listed therefore as a trading account debit. From the imbalance of this second account you could instantly see how your business was doing. You then rebalanced *that* account by transferring its profits to a third, capital, account.

The role that zero plays here accords with the persona it put on after failed apocalypses: a balance-point between negative and positive amounts, as there between past and future time. Together these new ways of thinking made negative numbers as real as their positive counterparts; and the friction of these mutually exclusive bodies, in turn, redefined zero.

But this was just the beginning of a new beginning. It set up a vocabulary for speaking of your dealings with the world in terms of transactions: exchanges where every action of yours affects others, and theirs, you. So Mattäus Schwartz, the bookkeeper of Jakob Fugger the Rich, wrote in his handbook of accounting in 1518 that double-entry was "a mirror in which you see both yourself and others, questions and answers." Now, once a style of speaking is established, it allows and in fact encourages you to imagine in ways that weren't available before. Didn't this new vocabulary lead in time to the framing in physics of its conservation laws: matter, momentum, energy neither created nor destroyed but exchanged - and to such insights as Newton's Third Law of Motion: for every action there is an equal and opposite reaction?

And again: the only way double entry bookkeeping can work, as Luca Pacioli made clear in the great summary he published in Venice in 1494, is by assigning numerical values to *everything*, even such intangibles as bad debt and good will, depreciation and reputation. A slow-working effect of this is to undermine the Great Paradigm: for while the real goods of the trading account are now joined by these more ethereal, at times even fictitious, bodies (so that you would think this was but one more instance of Here pointing to There), the assignment of monetary values to them all cannot but put them on a par, returning us toward the pre-paradigm days when the B of "A is like B" cast light back on A. *Value* becomes the leveller, whether in the eventual guise of dialectic materialism or the reduction of knowledge to quantification. At the height of the Great Paradigm, numbers were 'figures' but zero was a 'sign'. Now they are all figures.

The Arabic numerals advanced here and there along a very contorted front. The typical merchant's accountant did his arithmetic on a counting board but transferred the results to his ledgers in Roman numerals or words. The new symbols slowly ousted both: you find them creeping tentatively into the accounts of that redoubtable merchant from Prato, Francisco Datini, in 1366, only to be elbowed aside again and again by sturdy hulks of narrative. They hung around the back door, carrying entries from invoices to the account books, but there the old guard took over and put things in the old way. By the time of Luca Pacioli, Roman numerals were used mainly for dates and putting a stamp of solemnity on a document - but the keeping of sums was a different matter from how they were arrived at.

The rival camps of abacists with their boards and counters and algorists with their numerals must have formed up very early on, if a popular German ballad from the twelve hundreds is any evidence:

Nun ist auch hi gesundert
Lot vurste von Norwege
Ichn weyz, mit we vil hundert,
Ob Algorismus noch lebens plege
Unde Abakuc de geometrien kunde,
De heten vil tzo scaffen
Solten se ir allen tzal da haben funden.
Now here too is seen
Lot, Prince of Norway
With how many hundreds I know not.
If Algorismus were still alive
And Abakuc [= Abacus], skilled in geometry,
They would have much to do
To count all those they found.

There were half-hearted attempts at combining the two. Gerbert's counters with numbers on them, you recall, made a remarkably long-lived holy nonsense of both (abacists were called gerbertistas into the 12th century). In France for a time the lines on the counting board were replaced by special, unmarked counters signalling place-value (did these, or Gerbert's apices, lead the French versifier astray with his jeton worth zero?), while in England counters came to be bunched under columns for pounds, shillings and pence in ways that displayed their worth: a single marker above and to the left of the rest meant 10, to the right 5 (except under pence, of course, where it meant 12 - or 6). Perhaps you could also read off from these arrangements the degree of fondness for the eccentric.

As the battle heated up the shrewder writers placed their bets both ways. Ulrich Wagner published the first arithmetic in Germany in 1493, and he taught "calculation on the lines and with numerals." This became "on the lines or with quills" in the flood of books that followed, as in the Introduction for to Lerne to Recken with the Pen or with Counters published in St. Albans in 1537. But what was a father to do, send his child to the abacist or the algorist? The title page of Adam Riese's second book of computations, which appeared in 1529, shows a prospective customer looking dubiously from one to the other. Inside, Riese comments:

> I have found, in teaching young people, that
> it was always those who begin on the lines
> who are more adept and quicker than those
> who work with ciphers and a pen. On the lines
> they get finished counting... and stand on
> firmer ground. They may therefore with but
> little trouble complete their calculation
> with ciphers.

It must have been very like the perplexity parents face now in trying to decide whether their children should learn to read by phonics or word recognition.

By 1535 a German woodcut showing a man apprenticing a boy to an abacist is used to illustrate the sentence: "When such a guardian does wrong, he steals from his ward as well." At about the same time in England a certain John Palegrave announced that he could calculate six times faster with ciphers than "you can caste it ones by counters." Although the abacists fought a long rear-guard action, zero had defeated them by the time the 15th century had turned to the 16th. You see the triumph of the algorists in a woodcut from Gregor Reisch's *Margarita*

Arithmetic defeats the counting board

Philosophica, printed in 1503. Here the spirit of Arithmetic smiles on Boethius - believed at the time to be the inventor of the numerals - and Boethius smiles too, his right hand ready to continue reckoning as it points to a 0 on his table. But Pythagoras - representing the abacists - sits discomfited at his board, apparently still multiplying 1421 by 2, while Boethius flies through a much more involved calculation.

I recognise these two figures in the wonderfully metamorphosed setting of Dürer's paired engravings from 1514: "St. Jerome" and "Melencholia". Their expressions are the same, the lighting, the saint's table, the shallow perspective too (even, you might add, the floating figure bearing its legend). And we know that Dürer was familiar with Reisch's book, since as the art historian Erwin Panofsky points out, the paraphernalia surrounding the figure of Melencholia will be found in another woodcut from it: that of the Typus Geometriae. What did Dürer intend? Wasn't he contrasting the pagan world's frustration (Reisch's Pythagoras become the geometrising Melencholia) with the content that Christianity brought: St. Jerome for the Christian martyr Boethius?

Yet Adam Riese was right: people skilled in arithmetic on an abacus, or with their fingers, calculate much more quickly than do those who conquered them (excepting the rare John Palegraves) with the quill. What then is the significance of their victory? It is the question we asked long ago about the divergence of body and mind. Wordless manipulation will carry you with dash and glory to the outermost edges of arithmetic - but it will leave you stranded once you cross the border into algebra, and all the lands of mathematics that lie beyond. There thought travels by signs laced into a language that can speak even about itself; that lifts forms away from the substances they restrained, becoming abstractly substantial too. It is a language that lets us dwell on relations, playing a cool light over our wordless doings and so picking out their permanence.

This language came into its own when zero entered it as the sign for an operation: the operation of changing a digit's value by shifting its place. This opened the door to putting on a par the signs for quantities and for the operations on them, all subject now to yet more abstract operations on them, and those in turn to others, endlessly, each bearing the peculiar, defining mark of this language: that no matter where in its hierarchy an operation or relation stood, it was expressed by a sign of like status with the rest, in the matrix of their common grammar. These signs could be read as pointing beyond themselves to the language of which they were the parts.

10

ENTERTAINING
ANGELS

1. THE POWER OF NOTHING

Not only Baucis and Philemon, that pious old couple from Ovid's tale; not only the venerable Abraham and Sarah; but each of us, singly and together, has entertained an angel unawares. We just don't know who the stranger at the door might be. We just don't know, of the thousands of signals each hour that go hurrying past our mind's window, of the thousand signs we scan and skim and ignore, which ones may be compact of enormous power to unravel, which ones may point to the focusing of our scattered lights.

Zero tripped carelessly into the Renaissance with the Arabic numerals and made itself indispensible to our reckoning. But like all the magical helpers in the stories, so self-effacing was it, so unobtrusively did it neaten our middens away, that we paid it little heed and less respect. Once it was a number like other numbers, it was set to sweep the hearth while we went dancing.

You know what happens in the stories when the grand folk return: the elves scurry away or stiffen into invisibility, yet always seem to leave some disquieting trace of mischief. Look what we find. We thought we had settled in Chapter Seven that a/0 doesn't mean a thing since it means too much: it can be any number whatsoever. But is that still true if a is also 0? Is 0/0 always as meaningless as, say, 4/0 or -81/0? Might there be circumstances under which other numbers peer, like eyes, through the sideways mask 0/0 resembles? A revolution in all our thinking about mathematics is stirring behind the bold Renaissance façades. When it erupts our little doubts may be swept away - or may become great new certainties.

But until this backstage stirring forces itself on us, let us enjoy the festival of confident invention that in italy, Germany, England and France extended the discoveries the Indian mathematicians had made. They had clarified once and for all how zero behaves when coupled through adding, subtracting and multiplying; and now we understood the ways it goes mad when forced to divide. If we couldn't yet settle the little matter of 0/0, why not ask instead about zero's behavior among the exponents? These more sophisticated interactions should make as much sense for zero as for any other number. 5^7 means $5 \cdot 5 \cdot 5 \cdot 5 \cdot 5 \cdot 5 \cdot 5$, or 78,125; and 7^5 means $7 \cdot 7 \cdot 7 \cdot 7 \cdot 7$, which is 16,807. Raising to powers is fancy multiplication, just as multiplication is fancy addition.

There is no problem if we raise zero to some other power: 0^5 just means $0 \cdot 0 \cdot 0 \cdot 0 \cdot 0$, which is emphatically 0. But reverse these roles and ask: what does 5^0 mean? If you try to philosophise your way around this question you get caught up in terrible verbal cat's-cradles: is it 5 times itself no times, and if so, is that 0 or is it meaningless? Since $5^1 =$ 5, is 5^0 five less, so 0 again? But then what would 5^{-1} mean: 5 less than that, so -5? That doesn't sound likely. Or does 5^0 mean you aren't raising 5 to any power at all, so it is just 5? yet since $5^1 = 5$, that would lead to the impossible 1 = 0.

The mathematician's way out of this labyrinth is to put a hand firmly on the wall closing you in and follow wherever your steps now take you. We understand what 5^7 means.

We understand 5^4. Do we understand equally well $5^7 \cdot 5^4$? Of course: that's a string of 11 5s multiplied together: $7 + 4$ of them. And $5^7/5^4$? Just write it out:

$$\frac{5^7}{5^4} = \frac{5 \cdot 5 \cdot 5 \cdot 5 \cdot 5 \cdot 5 \cdot 5}{5 \cdot 5 \cdot 5 \cdot 5}$$

Is there any way of simplifying this fraction, short of doing all the multiplication out and then dividing? Yes: since $5/5$ is just 1 in disguise, we have four pairs of $5/5$ here, so $1 \cdot 1 \cdot 1 \cdot 1$, with $5 \cdot 5 \cdot 5$ left over. In other words,

$$\frac{5^7}{5^4} = 5^3.$$

It is as simple as that: in $5^7 \cdot 5^4$ you *add* the exponents to get 5^{11}, in $5^7/5^4$ you *subtract* them, $7 - 4$, to get 5^3.

This was the clue we needed, and following it we return to the minotaur 5^0 at the labyrinth's heart. 0 is any number - such as 7 - minus itself. So $5^7/5^7 = 5^{7-7} = 5^0$, and $5^7/5^7$ is just 1: hence $5^0 = 1$.

Since there is nothing special about 5, this rule must work universally: $a^0 = 1$ for any a. This result may be strange and perhaps unexpected, but it is a sure one.

Yet now I hear a voice as from a spectator at an epiphany. "For *any* a?" it asks. "What if $a = 0$: is 0^0 also equal to 1?" Unfortunately we can't use our new device, because $0^3 = 0/0 \cdot 0/0 \cdot 0/0$ and each $0/0$ alas, returns us to the issue we have put on hold.

Zero, like Old Chaos, is loose again but larger now in this land of exponents. It looks more like one of those Cape Breton toughs at a dance turned brawl: "Judique on the floor! Who'll put him off?" Let us try. If exponents got us into this trouble perhaps they can back us out of it. What are they, after all, but a notation intended to help us? They facilitate and evoke; their meaning flexibly stretches (as we just saw) to something beyond the counting numbers that our old acquaintance Diophantus invented them for.

The wonderful thing about exponents is that when we playfully extend their meaning it turns out we are forced to do so in only one way, if the new uses are to be consistent with the old. True, they are our invention and we have free will: but only to act compatibly with the world we've made. This is one of the great insights from mathematics into the human condition.

We will try to understand what 0^0 is by extending our exponents to still more general numbers, like negatives and fractions. Our insight about subtracting exponents tells us that $5^2/5^3 = 5^{2-3} = 5^{-1}$. On the other hand $5^2/5^3 = 5 \cdot 5/5 \cdot 5 \cdot 5$ $5 \lozenge 5$ which is $1/5$ once you turn each $5/5$ into 1. We *had* to define 5^{-1} as $1/5$, and similarly 5^{-2} as $1/5^2$, and so on: for just about any a, $a^{-n} = 1/a^n$.

And fractional exponents, like $1/2$? Remember that when you multiply $5^3 \cdot 5^4$ their exponents add, yielding 5^7. So $5^{1/2} \cdot 5^{1/2} = 5^{1/2+1/2} = 5^1$. But that must mean that $5^{1/2}$ is a number which *times itself* is 5 - and the only such number is $\sqrt{5}$. So $5^{1/2} = \sqrt{5}$. Likewise $5^{1/3} = \sqrt[3]{5}$, and so on. Thus armed, let's try to put 0^0 off the floor.

$0^3 = 0$, you must agree (since $0^3 = 0 \lozenge 0 \lozenge 0$), and $0^2 = 0$, and $0^1 = 0$. Now we've just found out that $0^{1/2} = \sqrt{0}$, and Bhaskara showed us that $\sqrt{0} = 0$. Likewise $0^{1/3} = \sqrt[3]{0}$ is 0, and so is $0^{1/4}$, $0^{1/5}$ etc. Insinuating yourself toward 0^0 in this way (by letting the exponent slither down toward 0), what could be more convincing than the claim that $0^0 = 0$?

What could be more - or equally - convincing is this. 5^0, we proved to ourselves, is 1. So is 4^0 and 3^0 and 2^0: they are each 1. So is 1^0. In fact $(1/2)^0$ must also be 1, and so must $(1/3)^0$, $(1/4)^0$ and so on. So if you creep toward 0^0 in *this* way, always keeping the 0 exponent but letting the base shrink down to 0, it is painfully obvious that 0^0 is 1.

What are we to do: is it 0 or 1, or both, or neither? The spirits of those who contributed to the invention of exponents crowd around: there is Nicole Oresme, the Bishop of Normandy, who dreamt up fractional exponents some time around 1360 - but not the exponent 0. There, from a hundred years later, is the physician Nicolas Chuquet, whom we met in the last chapter. He came up with a^0 but not with fractional exponents. The two are shaking their heads as they explain their devices to each other, for they cannot do anything with the cocky figure laughing up at them. There is the Lutheran minister Michael Stifel: he became a mathematician from trying to unravel the numbers in <u>Revelations</u> and <u>Daniel</u>, and sixty years after Chuquet saw how to use both zero and negative exponents. Perhaps he will know what to do - but all we hear him say is: "I might write a whole book concerning the marvellous things relating to numbers, but I must refrain and leave these things with eyes closed."

We had put out a bowl of milk for the little Robin Goodfellow of zero and here he is grown burly. Is he about to reveal himself as the bat-winged destroyer of our reason? No - but a mighty transformation is at work. Zero had long been developing into a kind of number giving value to other numbers. Now, under the incessant drive of mathematics toward the abstract and the general, it is going to turn as well into a kind of knowledge, giving value to other knowledge. What kind of knowledge will it be? Zero knowledge, of course. To play its part it will first have to go in disguise.

2. KNOWING SQUAT

These disguises of zero develop in settings we haven't yet considered: settings that probe properties of number in general. Because mathematics is an art, its makers enjoy creating new scenarios for it to perform in - just as novelists put their characters in confined situations and let the way people tend to behave unfold the plot (since character is fate). This is less artificial than it sounds in view of our natural situations being constricted: you have only to look at the engrossed variety of lives within a tree-stump to recognise how locally life is lived. You might think, however, that mathematics, like astronomy, moved through inhuman dimensions, with its gigantic quantities and endless progressions of numbers.

Yet picture the soap-bubbles a child floats on the summer air. Small and large, each is a perfect world, with colors sliding like continents over its surface. A mathematician can make such bubble-worlds too: little universes of numbers that run up to a point and then cycle back again. What if, for example, you shrink the infinite universe down to just 0 through 11: will these figures now act fantastically, or will the deep truths about number show themselves in this microcosm too? Take addition: $2 + 3$ will still be 5 and $1 + 8$ will be 9; but $6 + 7$? That can't be 13 because there is no 13 in this world. $6 + 7$ will be 1 again, since 13 coincides with 1 as we cycle around. It is as if we were doing arithmetic on a clock-face with 0 through 11 spaced evenly around it. And this *is* our familiar clock if we just shift the numbers up by one unit, counting from 1 to 12. But that means that in this doll-house world, 12 plays the role of 0: add it to any number and you simply get that number back again (twelve hours after 3 o'clock is 3 o'clock; and $11 + 12 = 11$).

If this strikes you as too trivial a game to play, notice that by taking it seriously, those astonishing people, rapid calculators, can tell you in an instant what time it will be in a million hours. For if it is now, say, 10 a. m., those million hours will consist of heaped-up 12s that change nothing, plus whatever heap is left that falls short of 12: and that is the time it will be. Divide a million by 12, that is, and the remainder will tell you the answer. Since the remainder is 4, it will be four hours past 10 a. m., or 2 p. m. And if carrying out this division in your head seems itself no small feat, observe that 12 into 100 leaves a remainder of 4, and as you continue the division, bringing down successive zeros, the remainder will be 4 again and again: we could have known the answer would be 10 + 4 = 14 o'clock, or 2 p. m., after our very first division.

What day of the week will it be ten million days from now, if today is Tuesday? In the world this question conjures up, 7 is 0, returning us to the day we started with. It is as if the days ticked off on a clock-face with the numbers 0 through 6. And again, divide 10,000,000 by 7. The quotient doesn't matter but the remainder does, and it is 3 (the same as the remainder left on dividing 10 by 7): three days after Tuesday is Friday, which is the answer.

French mathematicians in the 17th century were among the first Europeans to form and explore such glistening bubbles. The Maya, however, long anticipated them, since they probably managed their complicated calendars by looking only at remainders after dividing huge spans of time by 13 or 20, or the other lengths of their basic periods. All of these devisings have a biological basis, since our internal clocks operate, when left alone, on circadian rhythms, and each of the finely-tuned instruments of the living world around us takes its own number for zero - yet still these differently-spinning gears sufficiently mesh for the vast assemblage to continue and even to evolve.

These miniaturised worlds vividly show how art abstracts from life and mathematics abstracts from art. In the two centuries before the French began articulating their luminous globes, German and Dutch woodcarvers had made exquisite tiny landscapes out of boxwood: Lot and his daughters; an elaborate hunting scene with boars and rabbits; the Queen of Sheba visiting King Solomon - each in a hand's breadth. In these nutshell *numerical* landscapes any whole number can be induced to play the role of zero, and by doing so, give us the answers to questions about phenomena that repeat.

It would be wonderful if there were important structural similarities among all these universes that pulse to different cadences. They will materialise when we look again at exponents and see the surprising way they behave under these circumstances: ways that lie, for example, at the heart of the latest cryptography. For now our travels are about to take us from knowledge of zero to zero knowledge.

Think of the seven-day clock we just looked at, with its units marked 0 through 6. Choose from them whichever positive number you like (as card-sharps say) - for example, 3 - and raise it to the 6th power. $3^6 = 729$. Now subtract 1 from that, giving you 728. If you divide 728 by 7, you get no remainder: $3^6 - 1$ is the same as 0 on this clock. Try another of these numbers, such as 2. $2^6 - 1 = 63$, which again is the same as zero in this system. You'll get the same result with raising 1, 4, 5 or 6 to the sixth power and then subtracting one.

Is this a peculiarity of having worked with a seven-day clock, or of the number 6 (= 7 -1) we used as our exponent? Remarkably enough the answer is no. If you work with a clock of 5 numbers (0 through 4), each of the positive ones raised to the fourth power, with 1 subtracted, will be 0 on it, such as $3^4 - 1$ which is 80, and leaves no remainder on division by 5. Why stop here? A clock of the 19 numbers 0 through 18 yields 0 when you subtract one from any of its positive entries, raised to the 18th power, and divide the result by 19 (so $2^{18} - 1 = 262,143$, which equals 19 x 13,797). I know without doing the calculation that $13^{22} - 1$ is exactly divisible by 23, and (if you want truly Archimedean immensity)

$(273,889,154,767,432)^{11,111,111,111,111,111,110} - 1$

is exactly divisible by 11,111,111,111,111,111,111,111. How can I be so sure? Because the French lawyer and amateur mathematician,

Pierre de Fermat, whose Last Theorem was recently so resoundingly proven, came up in 1640 with what is now called his "little theorem". It says something we can understand much more readily if we recognize that all that gallimaufry of numbers, 5, 7, 19, 23 and the monster above are *primes*: numbers that have no factors save 1 and themselves. Fermat guessed and then proved that any prime number - call it p - will exactly divide any whole number less than it (call it a), when a is raised to the p-1 power and then has 1 subtracted from it.

Put so, his insight sounds both too intricate and too disembodied. It might be a touch more vivid were we to say: no remainder is left when p divides $a^{p-1} - 1$. Or if that reminds you too much of a frustrating Zen koan, let's leave the riddle but keep the Zen quality of laughing at emptiness:

Fermat

In the world of p
You can't tell a^{p-1}, less one,
From nothing.

We seem to have gone from the days when our symbol
for zero stood for all sorts of different numbers, to all sorts
of different numbers now actually standing for zero, each
in its own bubble. What is especially dazzling about Fermat's little theorem is that it not only reveals a common
feature of these worlds of cycle-length p, but does so in the
face of our appalling ignorance about prime numbers. We
are approaching zero knowledge: for given a prime, we
have no insight into how to produce or predict the next; we
know that they are vital to all of mathematics, but their
pattern (if they have one) continues to elude us, although
many a life-time has been devoted to seeking it out. If diddly squat is as close to squat as makes no nevermind, we
know diddly squat about these building-blocks of multiplication.

How could this ignorance increase our knowledge elsewhere? let's sample five different ways this has happened in
mathematics at various times. Make yourself comfortable
and have with your tasting a meringue, a dish of trifle or
any of the frothy confections they called in Elizabethan
times "empty dishes".

Uncork first the cryptographer's recent invention of an
almost unbreakable code that plays on our shared ignorance of the primes. The diabolically delightful trick here is
to publish to all and sundry what seem to be the very keys
for unlocking the code: two numbers, n and e. When your
agent in the Ministry wants to send you the specifications
of the torpedo guidance system, she simply uses n and e to
encode her message, and only you will be able to decode it.
Why is that? Because e depends in a roundabout way on n,
and n is the product of your own two secret and very large

primes, which we'll christen p and q ("very large", these days, comes to about 150 digits each). Anyone knowing them could unravel the message.

Why couldn't counterintelligence just set about factoring n? Because n is so big - some 300 digits long - that not even banks of the fastest computers available could break it down in time to catch the pair of you or know that the plans are gone.

There is one hitch, however, in all this scheming: as an old recipe for Rabbit Stew put it: "First catch your rabbit." To use this code you must likewise catch those two large primes, p and q. There are infinitely many primes but we know only a sprinkling of them - and a very thin sprinkle indeed in the range we want. But Fermat and his little theorem stand ready to help us. You know that if p is prime and a is any natural number less than p, then $a^{p-1} - 1$ is 0 in a cycle of length p. But this means that if $a^{p-1} - 1$ *isn't* 0 in such a cycle, p isn't prime! This gives you a method for finding the large prime you want: pick *any* sufficiently large number at random as a candidate for p, then choose an a such as 2, raise it to the p-1 power and then subtract 1. This is the sort of thing computers glory in, so let your computer do this work while you have another meringue. Let it also divide the result by p. If there is a remainder other than 0, then p couldn't have been prime, so you make a second choice (here too you might leave this choosing to your computer) and try again.

If there was no remainder, p is very likely prime: its chances of being composite are less than 1 in 10^{13}. If those odds aren't good enough for you, test p again with a different a, such as 3. The successful testimony of each witness a (showing that $a^{p-1} - 1$ is equivalent to 0) significantly betters your odds. Once satisfied, you find a q in the same way. Now you multiply p and q, keep them to yourself but publish n and the number e associated with it, and wait. The coded message that none but you can read is on its way, and the game (but not, you hope, the jig) is up.

Here is the second bottle. It holds a "zero-knowledge proof": a way of finding out whether someone is who he claims to be, although you yourself will remain ignorant of the right answer to the questions you ask him. Pause a moment to savor the bouquet of so absurd a situation. Now let's say (in the best traditions of the sedate mystery) that a plausible stranger shows up claiming he is the long-lost brother of the twins Ann and Anne. Having just been hired by them as their lawyer, you can't for the life of you tell them apart - but he should be able to. Very well. You sit him down in the over-stuffed parlor and ask one of the twins to come in. "Which is she?" you ask. He instantly says: "Ann." She confirms it. After she leaves you repeat the process. "Ann," he says, and again he is right. You keep this up, having one or the other come in at random, and time after time he guesses correctly. You still can't tell them apart, but after about thirty successful identifications you reckon that the odds are better than a billion to one that the fellow is plausible because he is no stranger. You reunite the family and pocket your fee, handed to you by Anne - or was it Ann? The moral of our allegory is that not only have you still no idea about which twin is which, but you know nothing whatever of how their brother could tell - nevertheless, you know he's The Real Thing.

Take down this third bottle: a jeroboam. It has to be large because it contains mathematical proofs that are too long or too intricate to check. If you wonder how such things could be - since what one human devises another can follow - the reason is that some recent proofs involve so many cases that only a computer could check them all: and then we're at the mercy of possible errors in the logic, the program or its execution. No matter: what one human can set in motion another, if sufficiently cunning, can oversee.

The cunning lies in first rewriting the purported proof so that an inconsistency anywhere in it will show up nearly everywhere, as if error propagated with all the vigor the Inquisition feared it had. Then you simply program your computer to take a few random samples of the rewritten proof. If no inconsistencies show up, the proof is almost certainly sound (the chances of mistaking an erroneous proof for a valid one are, says an expert, less than 1 in a quadrillion). In this way, the same expert casually declares, "a proof that would fill the entire known universe if written up with characters the size of a hydrogen atom could be verified on a SUN workstation in a matter of minutes." You couldn't ask for a more spectacular demonstration of mind's reach exceeding mechanism's grasp - at the cost (as the red lights flash) just of not knowing *what* the error is.

The fourth bottle is minute: even the aroma of the eszencia it contains makes you reel. This vintage was pressed from nothing but the assumption that something either is or isn't, and no third condition is imaginable. If you accept this "Law of the Excluded Middle", as Aristotle called it, you accept "proofs by contradiction", like the one you saw on pg. 75 that showed division by zero was impossible. Using this style of proof you can, for example, show that if you cram an infinite number of points into a small closed box, they may thin out here and there but at least one point will be so swarmed about by others that no matter how close you get to it, there will still be an infinite number of points around it. Where is this point? The proof cannot tell you. Why should the others accumulate here rather than there? Again, no help. In fact the proof only shows that this special point must exist and gives you not the faintest inkling of anything else about it. So it is always with proofs by contradiction, which turn as haughtily away as those assertions that there must be a god without letting you in on any of his properties. The situation is as unsatisfactory as stopping a stranger and asking him if he knows the time, and his answering: "Yes I do", and walking on. Some mathematicians object that you can't go around hoping to construct what deeply is by showing nothing about it save its logical necessity. Most, when they find themselves unable to build what they need, have reverted since ancient Greece to this sort of proof that at least it must exist, since if nohing else that gives them a kind of encouragement to go on looking for the body which cast this thin shadow.

The bottle tucked in the back, of such thick, dark glass that we can't even tell if it is full or empty, has a label written in the arcane script of mathematical logic. Here are the theorems the Austrian logician, Kurt Gödel, came up with more than half a century ago. They gave the sort of sickening spin to our knowing nothing that a curveball has as it drops away from the bat, leaving you flailing. For while we always knew that our certainties were a pool of light in the darkness, we believed as well that the pool had grown and was growing: that clarity would spread until obscurity was no more than a distant horizon.

Gödel (of whom the emblematic story is told that he never once said anything false, by dint of leaving unfinished any sentence he saw would miscarry - so that the body of his utterances remained consistent through being incomplete) proved - by an act of the most audacious and double-jointed mental athleticism - that there are statements whose truth, in a consistent formal system, cannot (by the very nature of proving) be decided there: within that system they can't be proven true or false. Curiously, the very manner in which such statements are shown to be undecidable shows too that they are in fact true. They could be proven, not in the system, but in a natural enough extension of it (though this would give rise to newly undecidable yet true statements, again provable in an extension of this new system -

ad infinitum). So a stalemated game of chess confirms the rules of chess; but here the moves and the rules are on a par.

Had Thomas Usk been walking along with us at our tasting, would he be murmuring now that knowing nothing has no might of signification of itself, yet gives power in signification to others?

3. THE FABRIC OF THIS VISION

These five excursions into the art of mathematics have been guided by the spirit of zero in the body of ignorance. While mathematics is first and foremost an art, it is an art applied by science to unriddle the world: and now zero will step out again as itself to lead this unriddling. All of our progress in physics, chemistry, biology, all of our triumphs in engineering and economics, come from expressing the hang of things in the language of shapes and numbers. Here in Cambridge, Massachusetts, where I live, i was once sitting by the Charles river on a fresh spring day. A white-haired old man paused and sat down on the bench beside me. "Do you see those arches?" he said, pointing to the Anderson Bridge. "I taught them for thirty years."

Those arches had been designed to bear stresses with grace and safety, and designing them meant first quantifying stress: part of the mathematising of nature and invention that began even before Luca Pacioli assigned numbers to everything imaginable. The way we subdue the world to our convenience, the way we catch its drift (grasping together the course of the planets and the arc of a stone) is through equations, those minimalist structures balanced as finely as a Calder mobile. In them we partner the unknown we want to pin down (how high can this building be, given its site? How cut your coat to suit your cloth?) with the constraints on it, represented by numbers.

Translating the jostle and clamor around us into equations is half the art: solving them the other. This is no piece of cake, and held for mathematicians in the past the same horrific fascination that it holds for anyone learning algebra today (only they had no teacher to help them, nor any answer book in the drawer). Solutions come from cajoling the equations, putting questions to them in the right way: reason's equivalent to "Open Sesame!" And it is here that zero plays its crucial role. For years Arabic mathematicians had teased their quadratics apart with clever schemes like "completing the square". In fact our word "algebra" comes from the title of Al-Khowarizmi's book <u>al-jabr</u> <u>wa'l muqabalah</u>, which I've seen translated as everything from "Restoration and Reduction" to "Completion and Comparison" ("Algebrista y Sangrador" was once also written over Spanish barber-shops, but there it meant Bonesetter and Bloodletter - more violence than is needed here).

How did al-jabr work? The method was so often illustrated with the equation

$$x^2 - 39 + 8x = -2x$$

that - in the words of one historian - it ran like a golden thread through four centuries of books from Al-Khowarizmi's in 825 on; and now through twelve. You first "restored" it (al-jabr) by moving the negative terms to the other side, making them positive: so $x^2 + 8x + 2x = 39$. Then you "reduced" it to $x^2 + 10x = 39$: that is, you combined like terms. Now you could use strategem and guile to compel the unknown, like Rumpelstiltskin, to tell you its name: in this case, 3 (and -13 too, if you could tolerate negative roots better than Bhaskara did).

The only problem was that you needed different methods for cases we would today never think of as distinct. Omar Khayyam has one for attacking this sort of equation,

Napier

whose form is $x^2 + px = q$; another for solving $x^2 + q = px$; and a third for $px + q = x^2$. We've seen before such a scatter of particulars eagerly waiting for a unifying insight: and we always expect it to come, being the children we are of the Great Paradigm shift, where scattered species are bound to be subsumed under a just if distant genus. We weren't to be disappointed, although it took a remarkable man, far away in the mists of Scotland, to see that by setting these and whole families of equations like them equal to zero, they could be solved with a single general technique.

This man was John Napier, Baron of Merchiston near Edinburgh.

When in the late 1500s his castle wasn't under siege and he wasn't fighting off raiders on the land and neighbors in the courts, he dabbled in the occult, designed "devises of sayling under the water" and "a round chariot of mettle made of the proofe of double muskett", bewitched pigeons, undertook to find buried treasure by sorcery, deduced - in 36 closely reasoned propositions of apocalyptic algebra -

the relation between prophetic and historical time, concluding that The Last Judgment would fall between 1688 and 1700; and invented logarithms. He was rumored by his neighbors, in the early years of the 17th century, to be in league with the Devil. His dressing all in black, with a jet-black cock as his constant companion, probably did little to dispel these rumors.

Napier's magic, however, was of a singularly crafty sort - perhaps because, as Sir Francis Bacon pointed out, histories make men wise; poets witty; but the mathematics subtle. So when it became clear that one of his servants was light-fingered, he assembled them all (as the story goes) and told them that his black cock would discover the thief. He chained it in a darkened room; each servant had singly to enter, stroke the cock and emerge. And he caught his man: the only one whose hands were clean, having been afraid to touch the bird which Napier had sprinkled with soot.

There was magic behind what Napier did with equations. He would take one that had some terms equal to some others and by repeated passes of al-jabr bring all the terms to the left, leaving only a zero on the right. These were what he called his "equations to nothing". Why was the trick so important? It depended on what at first seems rather inconsequential: if a product of two or more factors is equal to zero, then at least one of them must be zero as well. It actually helps your thinking to translate this into the language of mathematics:

If $ab = 0$ then a, b or both $= 0$.

You hasten to point out, however, that after his sleight-of-hand he hadn't products of factors but sums of terms with variables and coefficients in them, and a constant tacked on for good measure - so how can this fact apply, and what significance would it have even if it did?

Never underestimate a warlock. If you look at his turning Al-Khowarizmi's equation into $x^2 + 10x - 39 = 0$, you will see what Napier had in mind. What if we could rewrite that left side as a product? Well, we can, with a little adroit manoeuvering: $x^2 + 10x - 39$ is the same as $(x - 3)(x + 13)$. But if $(x - 3)(x + 13) = 0$, then one of those two factors must be zero: so either $x - 3 = 0$ or $x + 13 = 0$, and this tells you at once that $x = 3$ or $x = -13$.

This method always works when the x we are looking for is a rational mumber. In every case, once you find the factors that multiply together to give you the original "equation to nothing", you set each one in turn equal to zero and read off the resulting value of x that satisfies the equation. That's why our bridges stand up and our rockets fall down just where we want them to.

It took a touch of genius to think of using zero in this way: a way that anticipates the conservation laws of physics, by showing that all the vicissitudes the variable was subject to sum up to no change. How did Napier manage to come up with what we now take so dismissively for granted? "My ancestor lived on the borderline of fable and truth", said the Lord Napier of 1857. Perhaps imagination needs such shadowy surroundings for its tentative constructions to thrive.

Yet a serious problem with Napier's method remains. "Once you find the factors" - yes, but how find them? What are the shifts and feints making up that adroit manoeuvering? As well ask a T'ai Chi master for the computer print-out of his moves. There are some rules of thumb that help, some tricks of the trade - but you try to factor an expression you recognize again that mathematics is hazardous.

If you care to look more closely at these techniques lying somewhere between craft and art, you won't be surprised to find the trickster zero at play again and again, in yet another persona. Here is an oval window to look through: the vista will narrow away to more and more remote contrivings. In the foreground, at least, is a problem familiar from any first-year algebra course: solving $x^2 - 1 = 0$. If you set out your two hopeful parentheses with an x lodged at their fronts:

(x)(x) = 0,

it seems equally reasonable to put a 1 in the back of each. But $(x + 1)(x + 1) = 0$ yields $x^2 + 2x + 1 = 0$, with that extra 2x and the wrong sign for 1. If you try $(x - 1)(x - 1) = 0$ you get $x^2 - 2x + 1 = 0$, which is just differently bad. Since the only whole number you can put in each parenthesis clearly is 1, and you need to have those middle terms cancel out, the solution must be to have +1 in one factor, -1 in the other, since then you get

$(x + 1)(x - 1) = x^2 - x + x - 1 = 0$,

and -x + x turns into 0, giving you, as desired, the factored version of what you began with.

Notice that 0, disguised as -x + x, fleetingly appeared in order to do its job, then vanished. It doesn't even appear in the name that ennobles this technique, "The Difference of Two Squares" (in this case, x^2 and 1, which is the same as 1^2). This difference of two squares, $r^2 - s^2 = 0$, we now blithely affirm factors into

$(r + s)(r - s) = 0$. Half the potency of this magic lies in our demoting it so swiftly to the familiar.

Whenever you now see something like $x^2 - 64 = 0$ you factor it into

$(x + 8)(x - 8) = 0$. Even $x^4 - 64 = 0$ fits this pattern, since you can think of x^4 as $(x^2)^2$, so that $x^4 - 64 = 0$ factors into

$(x^2 + 8)(x^2 - 8) = 0$.

But what if, as occasionally happens, you want to factor an expression like $x^4 + 64 = 0$? Call up 0 again: but now its workings will seem more dazzling and farther removed from the ordinary. For $x^4 + 64 = 0$ looks intractable. The temptation is still to set up skeletal factors with x^2 in each: $(x^2)(x^2) = 0$ - but what do you do next? You must fill in the blanks with 8s to get 64, but now no combination of signs will succeed:

$(x^2 + 8)(x^2 + 8) = x^4 + 16x^2 + 64,$

$(x^2 - 8)(x^2 - 8) = x^4 - 16x^2 + 64,$

and

$(x^2 + 8)(x^2 - 8) = x^4 - 64.$

We're butting our heads against a sum, not a difference, of two squares: $(x^2)^2 + 8^2$.

For help, think back to the way we factored $x^2 - 1 = 0$. If you run what we did backward, you could say that we inserted 0 in the form of $-x + x$ between x^2 and -1: since then

$x^2 - x + x - 1 = 0$

factored into $(x + 1)(x - 1) = 0$. We need in the same way to insert 0 between x^4 and $+64$.

The extra subtlety now is to add and subtract a perfect square, hoping thus to end up with a *difference* of two squares, which we know we can factor. This strategy is the only trace of its nameless inventor. But what are the tactics for carrying it out? Another anonymous someone hit on the lucky combination: $16x^2 - 16x^2$. If we insert this into $x^4 + 64 = 0$ we get

$x^4 + 16x^2 - 16x^2 + 64 = 0$

That doesn't look particularly useful until experience suggests that you rearrange the terms like this:

$x^4 + 16x^2 + 64 - 16x^2 = 0.$

Add a pair of parentheses to see it:

$(x^4 + 16x^2 + 64) - 16x^2 = 0.$

We ran into that first term a minute ago: it is

$(x^2 + 8)(x^2 + 8),$

that is, the perfect square $(x^2 + 8)^2$. And

$16x^2 = (4x)^2.$

$x^2 + 8$ is the r, $4x$ the s, in $r^2 - s^2$, and since that factors into

$(r + s)(r - s),$

$(x^2 + 8)^2 - (4x)^2 = 0$

factors into

$(x^2 + 8 + 4x)(x^2 + 8 - 4x) = 0.$

That is how 0 allows us to factor

$x^4 + 64 = 0.$

Down the perspective lines from our oval window 0 diminishes away, lightly tapping apart ever more recalcitrant expressions.

$x^4 + x^2 + 1 = 0?$

Insert 0 in the form of $x^2 - x^2$, rearrange it this way:

$x^4 + 2x^2 + 1 - x^2 = 0$

which is

$(x^2 + 1)^2 - x^2 = 0$

and there stand its factors:

$(x^2 + x + 1)(x^2 - x + 1) = 0.$

And $x^5 + x + 1 = 0?$

Here add 0 in the monstrous form of

$(x^4 + x^3 + x^2) - (x^4 + x^3 + x^2)$

and you will come out, at last, with the original expression factored into

$(x^2 + x + 1)(x^3 - x^2 + 1) = 0.$

Almost out of sight now, zero, the choreographer of factoring, pirouettes on into calculus and then into the most inaccessible reaches of mathematics: number theory. Here it helped us to decipher the way things intractably are through the most delicate of touches. Taking on many disguises, it is itself imagination disguised. Are the facts that it has helped us understand therefore part fancy too? Perhaps Wiliam Blake was right when he said that "you clearly mistake when you say that visions of fancy are not to be found in the world. To me this world is all one continued vision of Fancy or Imagination."

4. LEAVING NO WRACK BEHIND

A difference between fancy and fact is that fancies may be as you please but facts are as the universe pleases. Where then does mathematics lie, that we build it as we will yet it dovetails with the world? Is it the mind's pia mater, a membrane through which the inner and outer exchange their currency? And do both somehow put their seals of approval on it, making mathematics more certain than anything else in our experience? For we take its conclusions neither on faith nor authority but as the last lines of proofs. These are at times as deceptively simple as a Chopin waltz, at times as monumental as a Beethoven quartet - but if they aren't musical they aren't mathematics.

Dr. Johnson once said that the good of counting was that it brought everything to a certainty which before floated in the mind indefinitely. But on what basis in turn do the laws of counting themselves rest? When we analysed just now the solving of equations, that complex business centered on the truth that if $ab = 0$ then a or b must be zero as well. And what does *that* truth spring from? Let's follow it down, not through time but through the explorations that have been broadly made, backing and filling. There may be some subterranean surprises.

We want to convince ourselves that if a isn't 0 but ab is, then b is 0. All the terrors of an equation can be driven away by the homely picture of a see-saw: supposing that $ab = 0$ means to imagine such a see-saw perfectly balanced, with ab on one side and 0 on the other.

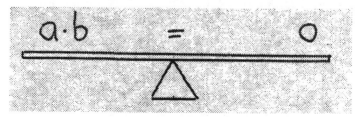

10.1

To keep this see-saw in trim, whatever you do to one end of it you must do equally to the other. We're supposing that a isn't 0, and want to prove that therefore b is. Since a isn't 0 we can divide by it - so go ahead and divide both sides by a, giving us

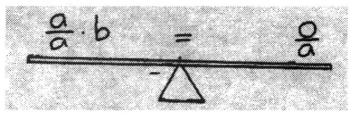

10.2

"a/a",we know, is just 1, so on the left side we really have 1◊b, and 1◊b is just b:

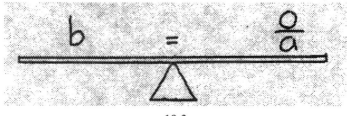

10.3

The last step is pleasant. "0/a" is our shorthand for (1/a)◊0. Since we're supposing a isn't 0, 1/a is some number. But any number times 0 is simply 0. So our final, balanced see-saw shows us that

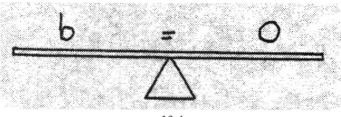

10.4

and since that is exactly what we had hoped for, we are done.

Or are we? A pursuit of truth that isn't relentless is no pursuit at all. I glibly said that $1/a \cdot 0 = 0$ because *any* number times 0 is 0. Why take that as a Commandment - can't we ask *why* it is true? Descend this staircase. We want to convince ourselves, on the solidest grounds, that $n \cdot 0 = 0$ for any number n (or a or k or whatever alias you choose for our anonymous plaintiff: we just want to keep things straight when speaking of "any number" in relation to any *other* number). Well, we know two much deeper truths. The first is that if you subtract any number at all from itself you are left with nothing: $k - k = 0$ for any number k. The other truth is about how multiplication and addition intertwine: if you multiply a number by the sum of two others, you get the same answer as if you had multiplied each term of the sum to start with by that number and then added the results:

$d \cdot (e + f) = d \cdot e + d \cdot f.$

This truth is called the Distributive Law, and is peculiar in being both fundamental and yet hard to remember and apply: children are always getting something like $5 \cdot (7 + 13)$ wrong, because while the answer is $5 \cdot 20 = 100$, which is the same as

$5 \cdot 7 + 5 \cdot 13,$

they often forget to multiply the 5 by one number or the other.

But we won't forget. We'll rub our two sticks of truth together to strike the spark that $n \cdot 0 = 0$. Since 0 and $k - k$ are the same, we can rewrite $n \cdot 0$ as $n(k - k)$. *Now* use the Distributive Law:

$n \cdot 0 = n \cdot (k - k) = n \cdot k - n \cdot k.$

But $n \cdot k$ is just some number, so $n \cdot k - n \cdot k$ is that number minus itself, which our Truth says is 0 - and there we are:

$n \cdot 0 = n \cdot (k - k) = n \cdot k - n \cdot k = 0.$

Cross all the bridges of equality and n◊0 stands on the opposite shore as 0.

Have we touched the ground zero of certainty at last? Are you convinced that in multiplication 0 is The Annihilator? A great insight into human nature surfaced from the purges of the French Revolution: "Un pur trouve toujours un plus pur qui l'épure": pure people always find someone yet purer to purify them. Don't we need to find even more fundamental truths to derive these last two "laws" from? And if we do, won't they require antecedents, and so down this eternal spiral to where their fires are not quenched? For a truth even deeper than the one which sprang from Robespierre and the revolutionary mob is that the kind of certainty demanded by deductive thought is unattainable because of the nature of deductive thought. To stop the infinite regress we have to say at some point: "We hold *these* truths to be *self-evident*".

Such are the "Distributive Law' and the truth that k - k = 0 for all numbers k. Call these ultimate stopping-places axioms, if you like (from the Greek for "what is thought worthy"); or postulates, with their Roman court-room air of being accepted only for the sake of the argument; or endow them with extra-deductive qualities like intuitive or tautological truth. Call them the arbitrary rules of a game we happen to be playing, or Revealed Truths, or (begging the issue) reflections of how our particular brains chance to work: these are all admissions that we have no higher court to appeal to: that *we* are the selves to whom these truths are evident.

Behind the fabric of our vision we are the scene-shifters and puppeteers: a disembodied company from the standpoint of the world's to-ing and fro-ing. This abstraction is an inevitable consequence of the Great Paradigm, which equates the ever more real with the ever more ineffable. And in this rarified atmosphere, zero will undergo yet another metamorphosis to suit itself to the ascetic life of axioms.

We have already seen zero becoming a number like other numbers as the Indian mathematicians shifted the emphasis from what they all were to what they did; then, as they became in common the values solved for in equations - so that their status changed to that of signs in a language that talked about structure. Numbers acquired a reality greater than that of mere things: we could speak not just of "four trees" but the unadorned four; not just of "zero thousands" but zero itself. Now, as we try to understand how all these numbers behave, we see (and say in our axioms) that their combinings matter more than they do: if we fully understood the operations of adding and multiplying, then the results of adding these numbers to those or multiplying them together would follow as surely as fruit matures in the summer sun.

The Distributive Axiom tells us how the operations of addition and multiplication interact. The other axioms we need developed from what I like to think of as a Newtonian point of view. For in thinking about gravitation, Newton decided to stop asking what it *is* (a fluid, a substance, a force?) and ask instead *how it behaves*. By shifting his focus from the medieval question to a much more abstract and dynamic one, he was able to discover that bodies, under gravity's influence, attracted one another inversely as the square of the distance between them. This proved in the end to be much more useful for understanding the world and predicting the positions of bodies in space.

In the same spirit, mathematicians eventually looked away from what addition and multiplication *were* and sat down to codify how they *acted*. Anxious lest they be misled by extraneous associations in the names and symbols for these operations, they spoke only of "an operation" on numbers, and denoted it by the neutral symbol "*". They eventually came up with these axioms:

1. If you do the operation * to any two numbers, a and b, you will produce another number, c. In short: a*b = c.

2. The order in which you operate on a and b doesn't matter, for the outcome will be the same. That is: a*b = b*a.

3. When you do the operation to three numbers, a, b and c, the outcome will be the same however you group them: a*(b*c) = (a*b)*c.

4. There is a special number - let's call it e - such that when you do the operation * on any number a and e, you just get a back again: a*e = a.

5. For any number a, there is another number - call it a' - such that when you do the operation * to these two, you produce the special number e: a*a' = e.

These axioms tell you everything about the operation *. But it was addition and multiplication, + and ◊, we were interested in. Which of these did * turn out to be? Strangely enough, it seems to describe both! Yet addition *isn't* multiplication: 2 + 3 ≠ 2·3. Have we not abstracted ourselves right past this crucial distinction?

This is where zero comes in to save the day. What is that special number, e, described in Axiom 4? For addition, e is 0: that is, a + 0 = a. For multiplication, however, e is 1: a·1 = a. 0 thus sheds all of the sacks of history loaded on its back to become addition's acolyte, *the additive identity*. Similarly 1 is estranged from the one that is one and all alone and evermore shall be so to become simply *the multi-*

plicative identity. In this return of the Bodhisattva ∗ to the world, we have to distinguish his two avatars and say: $0 \neq 1$.

This recognition forces a modification on axiom 5 as well, which is true as it stands for addition, where ∗ is + in disguise and e is 0: every number a has an *additive inverse,* a', which of course we write "-a", so that a∗a' = e turns into a + (-a) = 0. This axiom is the astral body of the insight first gained when Days of Judgment never came: then 0 stood as the present between the widening vistas of past and future time. The insight returned with double-entry bookkeeping, where 0 was the balancing-point of debit and credit. It is through this miraculous hoop of zero that positives turn to negatives, despite Winston Churchill's postprandial vision:

> I had a feeling once about Mathematics - that I
> saw it all. Depth beyond Depth was revealed to
> me - the Byss and the Abyss. I saw - as one might
> see the transit of Venus or even the Lord Mayor's
> Show - a quantity passing through infinity and
> changing its sign from plus to minus. I saw ex-
> actly how it happened and why the tergiversation
> was inevitable - but it was after dinner and I let
> it go.

For multiplication, however, Axiom 5 is no longer the whole truth. We have to say that every number *except* the additive identity, 0, has a *multiplicative inverse,* usually written 1/a, so that a∗a' = e, that is, a ·1/a = 1. Zero may be addition's acolyte, but multiplication preserves its renegade status.

The distributive axiom, by the way - with which we began this excursion - recognises in its asymmetry the distinction between addition and multiplication. It tells us that a·(b + c) = a·b + a·c. You can't, however, reverse the roles of + and · here: it isn't true that

a + (b·c) = (a + b)·(a + c).

The self in us that came trailing clouds of glory reached painfully back to collect or recollect these austere laws, in which 0 and 1 are the only actual numbers. From them (and an elusive last, yielding the irrationals) all the rest of the numbers in the world spring, and all their behavior is regulated by them. Who would go further back and question these laws must await, like the Spartans, their lawgiver's return, and abide by the promise to keep them until then. But Lycurgus never returned: that was his parting gift to Sparta.

II

ALMOST NOTHING

1. SLOUCHING TOWARD BETHLEHEM

Only selective forgetting of the past lets us move on, taking what was once dubious as the most banal of certainties, what was gained through struggle as our birthright. So with zero. The sermons it spoke in place-holding shrank to a letter of our thinking's alphabet, its volumes on solving equations to a sentence in mathematical primers.

By the seventeenth century our attitude toward equations themselves was changing. These snapshots of relations were yielding to functions - dynamic flip-books of the world - as our interest moved from how things are to how they become. The very nature of zero was about to undergo the most stunning transformation, as, in a different way, was the Great Paradigm which had supported its evolution over the preceding fifteen hundred years.

The problems coming into focus were problems of motion. How predict the course of a planet or perfect the flight of a shell? The difficulty was that both travelled along curved paths, while all our understanding, from Greek geometry on, was in terms of straight lines. If we knew, for example, what the slope of a curve was at a point on it, we could tell just where it was headed - but how could a curve even have a slope at one of its points? It might help to put the curve on a coordinate plane (which luckily had just been invented by Fermat - whom we've already met - and his contemporary, René Descartes)

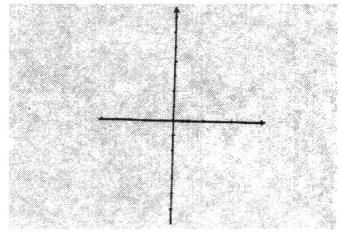

11.1

so that it would look like this:

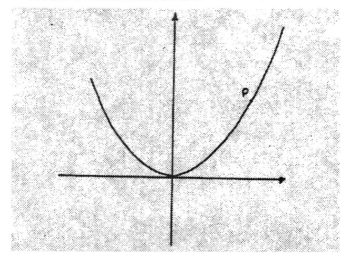

11.2

But slopes belong to straight lines, as here, where the path from A to P rises a vertical amount y while covering a horizontal distance x - and we

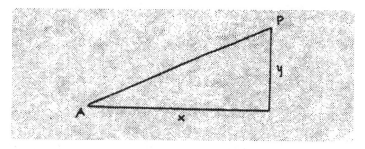

11.3

say its slope is the ratio y/x (so a 1 in 10 hill rises a unit for every ten it runs along: negligible in a car, exhausting for a cyclist).

Even to hazard a guess at what the slope of a smoothly changing curve at some point on it would be, as by drawing the line tangent to

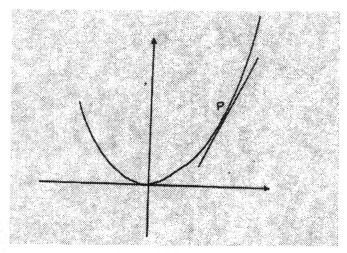

11.4

the curve there, is to raise the hackles of some and the laughter of others. As late as the 19th century Schopenhauer, pausing in mid-pessimism, cited the humor of this as evidence for his theory of the ludicrous (the joke seems to have been that there both is and isn't an angle here. It must have been the way he told it).

The astonishing resolution of this problem rested on thoughts planted long since: Greek roots that put out rootlets now and again in the middle ages; myths of the miniscule, revived and revised; a craftsman's appreciation of shims and splines; our growing adroitness with division and a care for the scraps of remainder it left in its wake. By the early years of the 17th century people began to toy with an idea broadly like this. The curve in question is given by some equation or formula. I'll use our modern way of speaking and modern notation to say that it is the graph of a function, $f(x)$, and I'll keep as a running exemplar the squaring function, $f(x) = x^2$. We want to find its slope at some point P on it.

That point P on the curve casts a shadow on the x-axis which I'll call r: a point which is r away from 0. The height of P above r is the same as the output of the function (whose path is the curve) when r is fed into it: in our example, P's height, $f(r)$, is r^2. If we now draw a line up from r to P, its length will be $f(r)$.

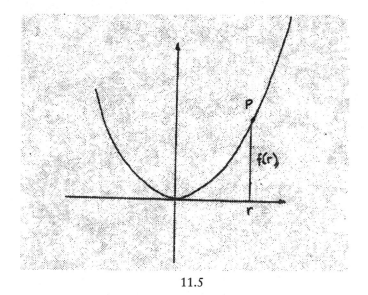

11.5

We now have the two coordinates of P: horizontally r, vertically f(r) (in our example, r and r^2).

Since the tangent is a straight line, two points determine it. We know that P is one of them. If we knew where the tangent crossed the x-axis - I'll call that point A - we would know the length of the line from A to r (call this length k), and so we would know the ratio f(r)/k, which is the slope of the tangent line to the curve at P: the slope we want.

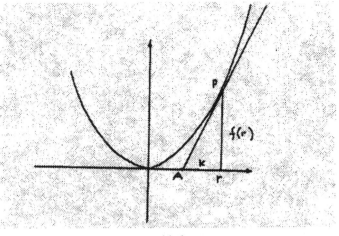

11.6

Notice that f(r) and k are the sides of a right triangle whose hypotenuse, AP, is part of our tangent. Here is where the magic begins. Choose any point on the x-axis a small distance h to the right of r - so this point will have the x-coordinate r + h - and draw a line vertically up from it, meeting the tangent at C and the curve itself at D. Draw too a horizontal line from P meeting this new vertical at B. Its length is also clearly h.

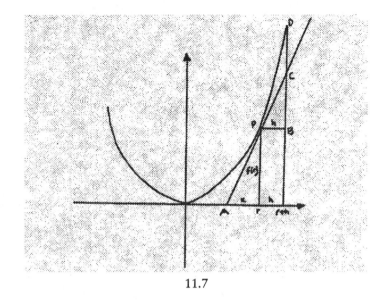

11.7

We don't know the y-coordinate of C but we do know that B's is the same as P's: $f(r)$; and D's is $f(r + h)$.

In our example,
$$f(r + h) = (r + h)^2$$
which is
$$r^2 + 2rh + h^2.$$
Why have we made this jungle-gym of new constructions? Because they make a triangle PBC similar to the triangle ArP, the ratio of whose sides, $f(r)/k$, we want to know.

Now similar triangles have been the stock-in-trade of geometers since long before Euclid, and they know that the ratios of the corresponding sides are the same. Here, therefore, $BC/h = f(r)/k$, so that if we only knew the length of BC we would be done. But of course we don't know it. We do know the length of BD, however: it is the length of the whole line-segment up from the x-axis, namely, f(r + h), less the distance up to B, which is f(r). In short, it is

f(r + h) - f(r).

Is this good enough? No, because C isn't D. But we can see that the ratio we want, f(r)/k, is the same as

$$\frac{f(r + h) - f(r)}{h}$$

So if we took our point r + h and began sliding it slowly left, toward r - shrinking h, that is - the vertical line would move with it, and look: the length CD would become smaller and smaller, until it finally disappeared! The triangle PBC and the shape PBD would become the same.

But here the magic runs into serious trouble. We have shrunk h

enough for it to disappear. That would make h = 0: and you know we can't divide by zero.

Mathematics makes the mind subtle. Focus its light on the squaring function we have used all along as our example, in hopes of finding a way ahead. In this particular case, the ratio

$$\frac{f(r + h) - f(r)}{h}$$

is

$$\frac{(r + h)^2 - r^2}{h}.$$

Since we can't let the h in the denominator be 0, let's actually carry out the squaring in the numerator first, and then cross our fingers. We get

$$\frac{r^2 + 2rh + h^2 - r^2}{h}.$$

Cancel the r^2 and $-r^2$, giving us

$$\frac{2rh + h^2}{h}.$$

Now, blindly following our luck, factor an h out of the numerator to get

$$\frac{h(2r + h)}{h}.$$

Miracle! Cancel those h's, leaving us with just

$$2r + h.$$

Now let the h shrink to 0 (Fermat, for one, simply says at this point: "remove it") and you are left with 2r. 2r is the slope of the tangent line to $f(x) = x^2$ at the point (r, r^2). So, for example, at (3, 9) the slope is $2\cdot3 = 6$. Try it: you'll find your measurements confirm our result.

Marvellous, revolutionary - and highly controversial. We can divide by 0 as long as we do so at the right moment? We can make logical sense or our sliding and shrinking, and of a triangle being the same, in the end, as a shape with a curved side? And what, precisely, does "in the end" mean?

Answers, objections and rebuttals that began in the late 17th century continued over the next hundred and fifty years, escalating to wrangles that passed from the polite to the bitter, growing from and contributing to the lore and language of the Almost Nothing. If the canny countryman knows that many a mickle maks a muckle; if all but invisible sprites and the Wee Folk have powers beyond ours; if less is more - why shouldn't least be most? Why not invoke the same trope here and rethink nothing as the trace of something's presence? This was the approach variously taken by one 17th century mathematician after another, rightly dazzled by the beckoning prospect of motion mastered. The story is too complex, both in mathematical and human terms, to do anything but tell it superficially here. Following Emerson's advice that when skating on thin ice, speed is your ally, I won't linger as long as I'd like, at the risk of blurring what was distinct and of leaving conclusions sparkling remotely, detached from the paths that lead to them.

It is certainly true that everyone who thought about it agreed that the curve and the line tangent to it grew closer and closer together as h grew very small. Since the problem lay in the difference between "very small" and zero (you can divide by the first but not the second), it became a matter for some of filling up the gap between the two with minute particles; for others, of animating this static picture.

Isaac Newton's teacher, Isaac Barrow, argued geometrically as we did just now (in fact our triangle PBC was long called "Barrow's differential triangle") and concluded that when this triangle is "sufficiently small" it might safely be identified with PBD. Geometrical methods, he said, were "freed from the loathsome burden of calculation." This is a telling comment, since calculation began with moving little pebbles, calculi, on a board, and what was developing here, which would go far beyond that, was *the calculus:* the movement of or through incomparably smaller pebbles, conceptual grains of sand, that would tell us what those calculi never could.

Yet how make sense of "sufficiently small"? As small as you like, said some, shifting the burden to the critic; "indefinitely" and "infinitely" small, said others, returning it to a world seen as composed of such atoms. Or were there atoms at all - might not everything be forever divisible? But an iota away from identity is still dissimilarity, and we want the slope of the curve not near but *at* P.

Hence in the camp of those who championed the notion of these "infinitesimals" a second tactic developed: the difference, or error, was less than any *assignable* quantity; or the very small could be ignored when compared to its larger neighbors (when we expanded $(r + h)^2$ and got $r^2 + 2rh + h^2$, we could have ignored h^2 as h grew small, because

h^2 grows smaller even faster). "Large quantities are like astronomical distances," said Johann Bernoulli in 1691, "and infinitely small ones are like the animalculae you see under a microscope" - so that "a quantity which is diminished or increased by an infinitely small quantity is neither increased nor decreased."

This calls up visions of the sequentially diminishing chests manufactured by Sergeant MacCruiskeen in Flann O'Brien's astonishing novel, The Third Policeman:

"Six years ago they began to get invisible, glass or no glass. Nobody has ever seen the last five I made because no glass is strong enough to make them big enough to be regarded truly as the smallest things ever made. Nobody can see me making them because my little tools are invisible into the same bargain. The one I am making now is nearly as small as nothing. Number One would hold a million of them at the same time and there would be room left over for a pair of woman's horse-breeches if they were rolled up. The dear knows where it will stop and terminate."

"Such work must be very hard on the eyes," I said.

The notion that almost nothing is tantamount to nothing at all stretches back at least to Meister Eckhart, if we are to believe his translator: "...[for] Eckhart between variable man and the constant God the difference may be made as small as you please." It is with us still, as in the urban legend of the bank-teller who grew rich by collecting each day the nugatory rounded-off thousandths from so many customers' accounts into his own. If a hockey goalie shuts out his opponents in six games out of seven, and has a scoreless streak of three hours, 51 minutes and 54 seconds in a season that lasts all winter, he is acclaimed Mr. Zero, as was Frank Brimsek, of the victorious Boston Bruins, in 1938.

Leibniz

One figure rose above all those who reasoned by infinitesimals: Gottfried Wilhelm Leibniz, born in 1646, the German diplomat, lawyer, philologist, philosopher, historian, geologist and co-founder of the calculus. At times he spoke of preserving the ratio of a triangle's sides even when those sides disappeared; at times he dropped negligibly small terms from his computations. He spoke of what he called dx as a difference in x-values between two points that were "infinitely near", but also described it as an "incipient quantity". His zeros, he said, weren't absolute but relative. On the other hand, he referred to them as a formal device with the same power as the imaginary numbers used so successfully in algebra.

You have to turn to his philosophy to come to grips with what this dx is: the answer is startling. Leibniz argued that the ultimate constituents of the universe can't be compound - if they were, they wouldn't be ultimate: you could break them into yet simpler parts. But any moment of time, any point of space, can always be subdivided, at least in thought. Each of us does, however, know an indivisible, ultimate entity: namely, the self. These are the *conceptual* or *metaphysical* points he called "monads", you inside yours and I am inside mine - or rather, each of us *is* a monad. You sense what he meant when you think of your unadorned, unconnected self: pure interiority. Perhaps this is why he spoke of the monads as "windowless." It is the monads, the indivisibly individual, that truly exists for Leibniz - and the general can only be thought.

And dx, the infinitesimal unit of mathematics? Like his monad, it too was active, not inert: each operated on the world, more verbs than nouns. He goes farther and calls mathematical points the "viewpoints" of his monads. For him the phenomena of the visible, spatial world were in effect translations of the monads' invisible structure. In light of the Great Paradigm, it seems to me, the monad was the afterimage of his dx, or what it foreshadowed. More, in the alphabet of concepts he devoted himself to inventing, dx may have stood for "monad".

How could so metaphysical a point of view - a point of view in which zero has turned into the innermost world of each of us - have led to a mathematics that so fully describes and predicts the external world? Because, Leibniz would have said, of the pre-established harmony of things. Out of harmony with his position, however, were those who held that motion could only be understood in terms of motion: a world in which not particles, no matter how small, but change itself was fundamental. And here that other founder of the calculus, Isaac Newton, stares distantly out at us.

In the sanitised, orthodox portrait he is the first thinker of the Age of Reason. Having read the papers Newton stored away, John Maynard Keynes came to see him as the last of the magicians, his deepest instincts esoteric, his occult goal to decode the secrets of the universe from the

Newton

starry cryptogram in which God had hidden them. The rival of Leibniz, he was nevertheless a living monad, secretive and inward - and only the narrowest and rarest windows opened into his works.

In 1665, when Newton was in his early twenties, the plague swept through London and his university, Cambridge. He hid himself away at his family's country house and set about solving the mysteries of motion. He was to say later that these were the two most fruitful years of his life. Like his teacher Barrow, he thought in terms of infinitesimals, lifting the symbol 'o' for them from a shrewd Scotsman, James Gregory. Since he used this 'o' the way we used h a few pages ago - adding, multiplying and dividing with it - how disconcerting to see his little zeros even in the denominators of ratios.

Once back in his rooms at Trinity College (where he lived, ironically, a secret Unitarian) - surrounded as chaotically as Dürer's Melencholia by his tools of astronomy and alchemy, his lenses, retorts, hermetic texts and calculations

- he slightly but significantly transmuted his terms. He still did what we saw done before: deleting (or as said, "expunging") negligibly small terms from his equations; but these he came to think of not as infinitely small quantities in space but as points of *time,* calling them "moments". Under the guise of removing the harshness (as he put it) from the doctrine of indivisibles, a more dynamic point of view was emerging. Did this happen continuously or in the twinkling of an eye? All we have are these little glimpes.

The last one is stunning: he rejects dropping terms involving his o, since "in mathematics," he says, "the minutest errors are not to be neglected." He drops instead the whole notion of infinitesimals, whether spatial or temporal: "I consider mathematical quantities... not as consisting of very small parts, but as described by a continuous motion." Curves, in fact, are generated by continuously moving points: "These geneses really take place in the nature of things, and are daily seen in the motion of bodies." Names he had previously invented for the variable *("fluent ")* and its changes *("fluxion"),* now, as his x's and y's flowed, came fully into their own. "Barrow's differential triangle" became "evanescent". Just as an error shows up everywhere in a proof made transparent, so does the underlying premise of a life seen from the right angle. Newton's insistence on asking not what something is but how it acts is all of a piece with his making motion fundamental.

In any case, here, matured, were the rival schools of calculus, drawn from the opposition we have been following throughout between zero conceived of as an object and as an action. Like the lovers in a story of Isak Dinesen's, they were two locked caskets, each holding the key to the other.

Neither side could measure its procedures by the standards of clarity set in Greek mathematics. Instead they fought among themselves and were bombarded by criticism on all sides. What else would you expect from trafficking with the notoriously wayward wee folk? The will o' the wisp in German is an Irrlicht (errant light) because it leads travellers astray. Bernard Nieuwentijdt, a Dutch physician, wrote in 1694 that he couldn't understand how the infinitely small differed from zero, or how a sum of infinitesimals could be finite. These methods, he said, led both to correct and absurd results; where they weren't obscure they were inconsistent, rejecting infinitely small quantities here but not there. In his famous pamphlet, The Analyst ("Addressed to an infidel mathematician"), Bishop Berkeley in 1734 said that these evanescent increments "...are neither finite quantities nor quantities infinitely small, nor yet nothing. May we not call them the ghosts of departed quantities?" And he asked "whether men may properly be said to proceed in a scientific method, without clearly conceiving the object they are conversant about, the end proposed, and the method by which it is pursued?"

The answers ranged from the end justifying the means (if these makeshifts served, well and good), and rigor being the concern not of mathematics but philosophy; to appeals to the heart, which like grace was superior to reason (besides, the paradoxes here were as useful as those in Christianity). "Allez en avant et la foi vous viendra", said the French mathematician d'Alembert: "Just go ahead and faith will follow." The language of rebuttal was at times parallel, at times skew, to the argument. So that strange genius, Blaise Pascal, asked that finesse rather than logic be applied to these processes, which as you saw are themselves made with finesse, refining rather than defining the ever finer. Leibniz, on the other hand, attacked Nieuwentijdt for being overly scrupulous: yet just such minutely troubling gravel was the substance of his approach.

These were explorers who felt their way without map or compass, and who wanted to be left to their tracking unflurried by having to file reports. All their images of areas made up of infinitely many indivisible lines, or forms remaining after their magnitudes had been removed, are telegrams from the interior, meant both to startle and silence the world. "I made it difficult," Newton told a friend, "to avoid being bated by little smatterers in mathematics."

It wasn't until the middle of the 19th century that a way of understanding developed in France and Germany which seemed at last satisfactory - a way that harkens back to the worn coin James Watt used to carry in his pocket to test the fit of pistons to their cylinders in his steam engines: the necessary gap had to be less than the thickness of his sixpence.

This understanding has an engineer's touch to it, characterised by a give and take of tolerance demanded and achieved. A dwindling sequence of numbers (such as the ratios of sides in our shrinking triangle) has a distinct limit if no matter how close you insist that they get to it, I can show that eventually they do by naming the term after which they all then lie that near or nearer. Is 1 the limit of $1/2 + 1/4 + 1/8 + 1/16 + ...$ and so on? Yes, if I can show that the sum gets as close to 1 as you like. Within a hundredth? Just add up the first seven terms and you'll find the series is only $1/128$ away from 1. Within a thousandth? Yes again: the first ten terms sum up to $1023/1024$, which is less than a thousandth away from 1.

Centuries went into the making of this criterion. A myriad myriad calculus students have sweated to grasp it and not a few of their teachers have shied away from this game of seeing and raising the stakes in precision. Yet for all its rigors and artifices, the idea of this game is caught in the story of the fine-drawn wire an American manufacturer sent to a rival in Germany at the end of the First World War, a proud token of what his country could now do. It came back with a hole drilled neatly through it.

The ancient notion of "limit", tightened and trimmed, is the key here: a fixed, real number that a sequence of numbers slouches toward (so in our example $f(x) = x^2$, 6 is the limit of those ever better approximates to the tangent's slope at (3,9)). The shrinking has to be continuous: it can't hop or skip over gaps on its way to this goal, and you must be able to shrink down to the limit from either side and get the same result (if a rising hill ends in a cliff, it wouldn't make much sense to talk about its slope right at the cliff's edge). Once these technicalities were attended to and the symbolism that expressed them made convenient, the mystery (and perhaps some of the charm) was gone from the notion of a smooth curve's slope.

You may have guessed that this wasn't the only time a procedure - such as this shrinking of h to 0 - has been used in mathematics before it was formally justified: it has happened over and over again, from Archimedes on, because it springs from the ever-present tension between intuition and proof. These are the two poles of all mathematical thought. The first centers the free play of mind, which browses on the pastures of phenomena and from its ruminations invents objects so beautiful in themselves, relations that work so elegantly, both fitting in so well with our other inventions and clarifying their surroundings, that world and mind stand revealed each as the other's invention, conformably with the unique way that Things Are.

After invention the second activity begins, passing from admiring to justifying the works of mind. Its pole is centered in the careful, artful deliberations which legalise those insights by deriving them, through a few deductive rules, from the Spartan core of axioms (a legal fiction or two may be invented along the way, but these will dwindle to zero once their facilitating is over). What emerges, safe from error and ambiguity, others in remote places and times may follow and fully understand.

Now a peculiarity of formalising some great coherent body of mathematical insights is that (as we came to see only in the twentieth century) the formal reworking - a piece of language - cannot have a unique referent: it can be represented in many essentially different ways (like those make-your-own robot toys that are alligators with machine-gun eyes one minute and reassemble to sleek butlers the next, all with a twist of a bolt or two and the deadpan instruction manual). This means that we cannot pick out the one and only way Things Are from the multitude of models: we might even conclude (by an abuse of language) that there may not *be* a unique way that things must Be. This relativising of our thought goes past the post of the most post-modern excesses.

Perhaps our way of formalising is flawed. It may be too limited. Falling into the trap of its own relativising, it may be only one of possibly many ways to legalise, so that another might better judge different replicas of the world. Yet "replica" suggests an original distinct from its copies, and we can make no such distinction here. Each of these models that satisfies the formal system is differently but equally ours. The body-snatchers have invaded, and one simulacrum is no more inimical than another.

There is no denying the great advantages that rich formal systems have to offer us. They chase false insights away and bring those that cohere with our axioms into a language where they can be spoken of with their peers. Here new linguistic structures may coalesce that will contribute to further insights. You could even go on to claim that its ambiguities are benign, since what one embodiment of the system conceals another may reveal, just as a problem may yield to algebra that resisted the geometer's blandishments.

Still we have to acknowledge that the disguises such a system must put on keeps us from seizing The Only Way Things Are. The sense of a single truth grew from the prior half of doing mathematics, grown apart from its legislative twin. There those insights were gained in ways that we cannot describe and indeed hardly grasp, *and which by their very nature cannot be formalised*. The answer to Bishop Berkeley's query is that in fact we do best proceed at first when our ends, objects and methods are still pliable, lolly-gagging around within the green precincts of certainty: for the idea of a single way that things are belongs to this morning half - so of course we cannot expect it to survive the serious afternoon. The world may not only be more singular than we think, it may be more singular than we *can* think.

2. TWO VICTORIES, A DEFEAT AND DISTANT THUNDER

What was born with the calculus wasn't only a way to catch and control the spectacle of change, but a new sense of where significance lay. Or was this sense not new but renewed, having awaited its renaissance a very long time? For the inklings of a shift in the Great Paradigm, which we first heard in the whispered exchanges of double-entry bookkeeping, now hummed in the sliding of a line down to tangency with a curve. The sound came from the notion "limit": an *action* (the process of shrinking) points to an *object* (this very limit) - but the components of the action and the object are all of a kind: numbers. We routinely go so far now as to signal this pointing by an equality sign, as if the going and the being there were the same. The limit lights up the process, revealing its shape: the B in "A is like B" explains A. Newton had to be the last of the magicians, because after him the seen no longer pointed to the unseen but (in ways that might take heroic eforts to uncover) back to itself.

The practitioners of this art built up and lived in the immense, ramshackle palace of calculus, even while the bricks of its foundation were being repointed and heftier joists fitted to take up the sag. From its heights one region after another of the physical world was surveyed and governed: optics, hydrodynamics, mechanics; and farther afield, biology, economics - all of the sciences, pure and applied. Even an old puzzle of ours fell to its power: for this was the revolution in our understanding of numbers which would (in the right setting) give the indeterminate 0/0 a definite meaning.

The story is one of disguises. Someone late in the 17th century - shall we call him Guillaume François Antoine, Marquis de l'Hôpital? - considered two smoothly changing functions, $f(x)$ and $g(x)$ (such as $f(x) = 2x$ and $g(x) = 3x$). Their ratio makes perfect sense for practically any x. In our case, if $x = 17$ then $f(17) = 2 \cdot 17 = 34$ and $g(17) = 3 \cdot 17 = 51$, so

$$\frac{f(x)}{g(x)} = \frac{34}{51} = \frac{2}{3}.$$

It will be 2/3 for just about any other x you choose. Just about any - but not $x = 0$, for then we have

$$\frac{f(0)}{g(0)} = \frac{2 \cdot 0}{3 \cdot 0} = \frac{0}{0},$$

our old nemesis.

What was noticed, however, was that if each of these functions independently had a slope at the critical place (in our example, at $x = 0$) - and if the slope of $g(x)$ wasn't zero there - then the ratio of their *slopes* was the same as the ratio of the functions themselves!

To put this in the language of limits and look at our example: since the limits, as x went to 0, of $f(x)$ and $g(x)$ were both 0 ($2x$ and $3x$ both go to 0 as x does); and since the slope of $f(x) = 2x$ is everywhere 2 and that of $g(x) = 3x$

is everywhere 3; the limit, as x goes to 0, of f(x)/g(x) is the same as the limit as x goes to 0 of their slopes: here, 2/3. The shorthand for the slope of a function is (shades of old Greek usages) a small vertical dash above it and to the right - so f'(x) reads off the slope of f(x) at the input x. We can therefore condense this revelation (with "lim" standing for limit) to the alchemical-looking

$$\lim_{x \to 0} \frac{f(x)}{g(x)} = \lim_{x \to 0} \frac{f'(x)}{g'(x)}$$

In other words, 0/0 was a disguise - in this case, for 2/3.

The Marquis was immortalised by his discovery of this general principle, which now goes under the name of l'Hôpital's Rule. The only problem with our story - which I told you was one of disguises - is that Monsieur le Marquis had neither this insight nor its proof. Both were the work of his teacher, Johann Bernoulli, who was apparently willing to take the Marquis' cash and let the credit go. Since it is infinitesimally more fun to speak of "l'Hôpital's Rule" than "Bernoulli's Rule", I doubt that justice will ever be fully done to the author of this notable victory.

The surrender of 0/0 is, remember, conditional: it is only in the context of slopes and their ratios that it has meaning. Otherwise division by zero remains forever impossible in our full-fledged world of numbers. This doesn't return us to the days when zero itself was questionable - unless you live in Pennsylvania. For in May 1998 the Commissioners of Centre County voted to abolish an unpopular tax on occupations by evaluating them all at zero. Since this left the local schoolboards painfully short of income, they sued the county, claiming through their lawyer that zero was not a value. His proof was to have a former county assessor try to divide by zero on his pocket calculator. Only the 'E' for 'error' appeared.

For the rest of us, zero is - or has - a value. Whether it is more an action than an object is the Elusive She we have been pursuing through these pages. But can you *count* with zero? This must be equally elusive, since Dick Teresi, the science writer, recently called the Massachusetts Institute of Technology to ask just this of the people in its math department. He held on to the phone while the question reverberated up the corridors and down, until the word came back at last that no one could really say for sure; and anyway they were interested only in numbers invented after 1972, so he had better call Harvard.

Another question, you recall, has been waiting for something revolutionary to answer it: the question of whether 0^0 is 0 or 1 or what. From our new point of view we can summarise neatly what led to our two conjectures. We first looked at the limit, as x dwindled down to 0, of x^0, and convinced ourselves by sampling the process at way-stations that the answer had to be 1:

$$\lim_{x \to 0} x^0 = 1.$$

To be a touch more precise, that dwindling down was from larger numbers to smaller: we were taking the limit *from the right* as x went to 0, and we indicate this by writing: $x \to 0^+$. So our conclusion was that

$$\lim_{x \to 0^+} x^0 = 1.$$

We then took the limit (again as x went to 0 from the right) of 0^x, and again, by judicious sampling, saw it should be 0:

$$\lim_{x \to 0^+} 0^x = 0.$$

Each way of looking seemed equally sound, but one at least had to be wrong.

Perhaps the difference lay in the asymmetry of the set-up: in one the base, in the other the exponent it was raised to, was the variable. Why not let *both* vary, and take the limit as x goes to 0 from the right of x^x? By cleverly applying - what shall we call it, Bernoulli's Rule? - we find a definite answer:

$$\lim_{x \to 0^+} x^x = 1 \ !$$

Instead of working through the cleverness, let me just show you a graph of the function $f(x) = x^x$:

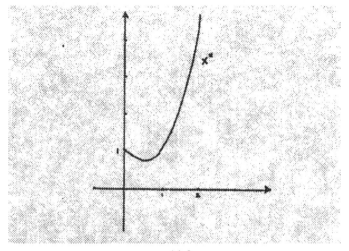

11.8

As you move along the x-axis from the right toward 0, sure enough the curve rises from its valley toward 1.

But stay your soul from clapping hands and singing just yet. You remember that a few pages back I said a limit had to be approached from either side to make sense - and here we seem to have indeed a rising hill ending abruptly in the cliff that I described there.

Well, you say, just complete the picture on the left of the y-axis, putting in negative values to x^x, and march through them to $x = 0$. I wish I could. Four distinct phenomena we've encountered on our travels here collide to make this impossible. First, we must find no gaps as we take our way: our graph must be continuous. But (second and third) look at what happens when you try to put in x = -1/2, for example, to the function $f(x) = x^x$. $(-1/2)^{-1/2}$ means (as we saw when we were entertaining angels) : $1/\sqrt{1/2}$, since the exponent 1/2 stands for taking square roots, and its negative sign forces us to put the result in the denominator. No harm having it there, if it were a real number - but what is $\sqrt{1/2}$? it is imaginary; and this fourth phenomenon means we do have a break and are whisked away from the real numbers to the broader, subtler plane that harbors imaginary numbers too: those numbers, like $\sqrt{1}$, symbolised by the letter i, which you will find nowhere among the reals but could picture as lying on an axis all their own and perpendicular to the real one.

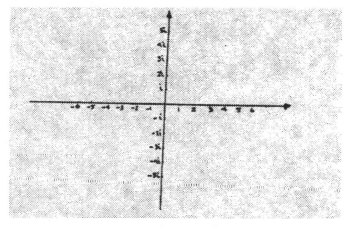

11.9

Here we can represent any combination of real plus imaginary, such as 3 + 4i, by a point on the plane with the appropriate horizontal and vertical coordinates (in this case, (3, 4)).

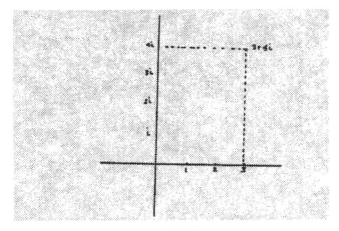

11.10

We want to find out what 0^0 is by slipping stealthily toward it. But now that we are on a plane rather than a line, we have to be sure that we can approach 0^0 from any direction. We've called our function $f(x) = x^x$ up to now, because the variable x stood for any real number. In order to take into account that these "complex numbers" are made up of real (x) *and* imaginary (yi) parts, let's change the name of our variable to z, where $z = x + yi$. We want to see what happens to $f(z) = z^z$ as z goes to 0 - that is, as x and y both go to 0.

What happens is monstrously strange. No matter along which path we approach (0, 0), the outputs of our function go haywire. They jitter through every number you can name, more and more chaotically the closer we get. They never settle down, they never converge to any particular value (much less 1). (0, 0) is the world's worst nightmare for z^z. It would be a colossal understatement to say it had no value whatever there.

We may be going down to defeat, but we'll go down with all flags flying. The design on one of them is a pattern of points in those lefthand quadrants, back on the real plane, where we *do* get real outputs from negative inputs to x^x. The situation is, as you might

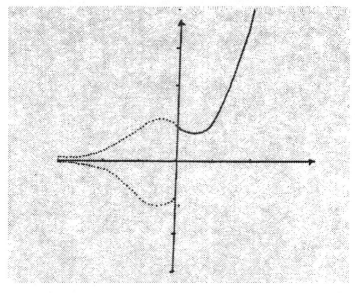

11.11

expect, very complicated: just as -2 is *a* cube root of -8, we will find real outputs here and there when x is a negative rational number with an odd denominator. We will, however, get multiple outputs for any input - such as two square roots, three cube roots, four fourth roots and more, with gaps opening up everywhere among the real outputs. But at least we can say this: while the function's graph, for these negative inputs, wraps round and round again, what it wraps around looks like a spindle - so there is some order in all this chaos.

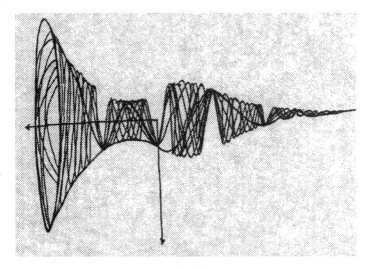

11.12

A second flag flutters in the breeze from a familiar quarter: the irrepressible urge in mathematics to generalize. If the function $f(x) = x^x$ has such interesting pathologies, what about a function x^{x^x}, or another, $x^{x^{x^x}}$, or longer and longer ladders of the variable raised to the variable raised to the variable... until you find yourself climbing the tower that Babel would have been? The answer (which comes from repeated applications of "l'Hopital's Rule") is predictably ironic: as x tends to 0 from the right, the limit is 0 if the number of x's in the tower is odd, 1 if it is even.

Of course this doesn't do anything for the hysterical ghost of 0^0: wish as we may, will as we might, the world's structure is too astonishing for it to materialise. But mind has its own marvels, and finds this way out of our dilemma. If you look at any polynomial you see that it ends with a constant term:

$$17x^3 - 8x + 3$$

or
$$102x^{19} - 14x^8 + 5x^5 - 7,$$
or even
$$x^2 + x$$
which has a tacit 0 at its end.

To honor this rationalised accounting, then (especially useful when we multiply polynomials together), it helps to arrange each one as here, with the powers of x descending from the largest; and to show the variable in every term. Where then is x^2 in our first example? Implicitly there, with 0 in its bookkeeping avatar to the rescue again, as coefficient. Had we written it out in full it would have been:
$$17x^3 + 0x^2 - 8x + 3.$$

And where is the variable in the final term of each? There again, but (conformably with the descent of powers) raised to the 0, since we know that $x^0 = 1$. So rewrite our example yet again as
$$17x^3 + 0x^2 - 8x + 3x^0.$$

The last term is indubitably 3 and must remain so whatever value x has - even should it take on the value 0. Treat the constant, therefore, as the coefficient of x^0, and stipulate that $x^0 = 1$ *whatever* x is: even when x = 0, $0^0 = 1$.

"Stipulating", and acting for the sake of widening the notation: we find ourselves uneasily with more leeway than when we extended our exponents from natural numbers to 0, negatives and fractions. *Then* we had to define the new usages in only one way if they were to fit in with the old. *Now* it appears that the convergence of two lines of thought in the symbol 0^0 has no intrinsic meaning, and we may therefore choose one for it on grammatical or aesthetic grounds.

This defeats the hope cherished by those such as Leibniz (so thoughtful about his notation) that the new linguistic structures born of formalising intuition will always yield new insights into the world this language speaks of. Apparently sign and referent are coupled more flexibly, and we make our agreements in the slender wedge between necessity and convention.

An instructive defeat, then, after a victory. But we now come to zero's greatest triumph in expanding our knowledge. Thanks to calculus, zero holds the key to our making the most of any undertaking, and doing so with the least effort. It holds the key as well to our understanding how things work, for while this may not be the best of all possible worlds, it is in its mechanics the optimal: best under given circumstances.

Why is zero the key? Because "Nothing takes place in the world whose meaning is not that of some maximum or minimum", as Leonhard Euler, the greatest mathematician of the decades after Newton, put it. The fall of light, the pricing of goods, a cheetah's turn of speed, the shape of a wing, a leaf, a mountain torrent, all the doubling back and going on, the minute and magnificent tinkerings, are each the solution to a problem of optimization. How can we predict when the process that concerns us will reach its desired extreme? By coming up with a smoothly changing function that describes this process, and graphing it; then noticing that at the crests and troughs of the graph the slope of the line tangent to it is - zero!

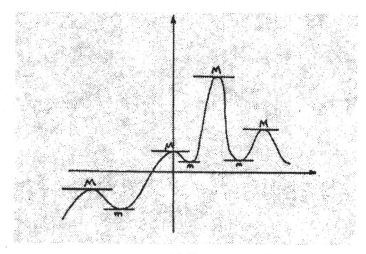

11.13

In fact we needn't even draw the graph; we derive another function from the first which reads off the slope of that function's graph at any point on it: the function f'(x) described before. Where this new, derived, function yields a zero, its input is the number which will maximise or minimise the process (a little care is needed to decide which: the maxima occur where - reading from left to right - the slopes turn from positive to negative, the minima where they turn from negative to positive. Then too, you might be deceived into thinking you had reached the mountain's top or the valley's floor when actually you were just poised on a ledge. The same sign to left and right of such a zero slope tells you to struggle on).

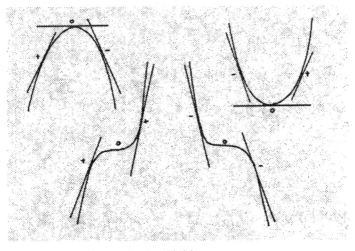

11.14

If there are several extreme points, we find the greatest or least just by comparing the original function's outputs at these points).

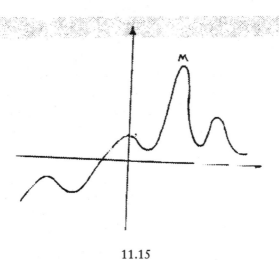

11.15

You would expect, in the case of zero, that what goes around comes around. Here the radical new conception, embodied in calculus, of zero far off as the limit approached by a receding sequence of numbers, has produced anything but so faint a persona: rather, the substantial zero which is our chaperone at the dance of change. It is no exaggeration to see all of engineering and science as devoted to eliciting and interpreting her signals, since if Euler is right, these point to the world's meaning.

How easy and exhilarating to write a victor's history, showing events having had to fall out as they did - a growing neglect of infinitesimals, the triumph of the view that made change fundamental, the 19th century work to legalise the notion of limit - since these maximise progress, and without them we would still be plunged in metaphysical speculations about worlds in a grain of sand. Yet once again we see da Vinci's scrawl bleeding through the glazes laid down by time: Tell me if ever anything was *done?* For we are always between two wars, and victories are only relative maxima. You can hear again, after so many years of shrinking, scorned and neglected, the rolling sounds of infinitesimals gathering, as their latest proponents mount a sophisticated attack on the castle Newton built.

These reborn infinitesimals lie at the intersection of two themes we have been following: zero as object and formalism as guarantor of validity. I mentioned that new insights might arise from the linguistic structures we make for putting the seal of approval on our understanding - and just this happened, beginning some fifty years ago, in the thought of a refugee from Germany, Abraham Robinson. For all that in California his body drove the latest model red sports car, his mind moved among models of a very different sort: those utter abstractions newly popular and powerful that established the consistency of a collection of axioms.

The idea was this. If you had a series of rules by which a clock, for example, was to work; and if when you tried building it you found that one rule required a lever to go down at the same instant that another rule demanded it go up, then you would see that your rules were inconsistent: the contradiction they led to meant that no model of them could be made. Hence if you could make or find a model of a set of rules, then that set must be consistent. Of course the model need not be made of cedarwood or even balsa and tissue paper: its parts might be wholly conceptual, its assembly the work of mind. Think of our axioms at the end of Chapter Ten, that legislated how adding and multiplying work: the integers are a model of them, and so are the rational numbers and the reals. The fact that there could be essentially different models of the same set of axioms - that fact we found troublesome a few pages ago, when it took away a unique view of the world - gave Robinson his clue.

What he did was to make a "non-standard" model of the axioms governing calculus (or analysis, as it is called in its upper reaches) containing all of the familiar real numbers but having as well some very peculiar numbers indeed: numbers greater than zero, yet less than any real number you could name. These truly almost nothings were his infinitesimals, and using them he and others proved all the traditional and some new theorems with an economy and ease that the cumbersome 19th century machinery could never manage. They restored honor to Leibniz and a granular basis to our thinking about change.

A final victory? The battle still rages, because if Leibniz' infinitesimals were ghosts of departed quantities, Robinson's seem to be ghosts of what aren't even quantities but linguistic expressions. They exist more in the language of formal logic than in the world, and even in that language they are closer to punctuation marks than to letters, syllables, words. Well, you might answer, our household zero too began life as a mere punctuator. True, and we cannot predict how these infinitesimals may evolve. Yet Robinson himself insisted on thinking of them not as real but as useful (or well-founded) fictions, and reminded us that Leibniz shared this view. He claimed, remarkably enough, that he had not invented new *objects* but only "new deductive procedures" (so that even within this camp actions were pointed to, but as ever in the recursive abstracting which is mathematics, actions on a remote and receding plane).

It is as if you had decided that the connectives of our language ("and", "or", "but", and so on) were *also* names of things, and in the fresh world enriched by these creatures, found that previously opaque notions suddenly became transparent. You might be loath to forego these insights just because somewhat dubious intermediaries brought them about. You might even begin to mull over the conceit that "but" and "or" have - somehow - as much of a claim to existence as the objects and actions whose names they stand between: a point of view which the German poet Christian Morgenstern in 1905 first looked from:

> One time there was a picket fence
> with space to gaze from hence to thence.
>
> An architect who saw this sight
> approached it suddenly one night,
>
> *removed the spaces* from the fence,
> and built of them a residence.

Whether you think of its airy walls as made of infinitesimals or motion, the house that zero has lived in since the invention of calculus is worlds away from the baked clay of its Sumerian origins. It would have been meaningless then to pose the question of our epigraph: how close is zero to zero? Now we ask it without flinching.

12

IS IT OUT THERE?

Nature abhors a vacuum and so do we. Zero, we've seen, is intricately woven into the workings of our thought, but the temptation time and again has been to look for its original outside of the mind, in physical space: a silent desert amid the clamor of oases. We may be disappointed. Like those cartographers in the Age of Discovery who filled up the vacant corners of their maps with winds and serpents, and wrote over still unexplored regions, "Here be Monsters", the cosmologists who are our modern navigators of stellar immensity posit brief surges of energy and fleeting bursts of quasi-quanta where you might have thought nothing was happening and there wasn't anything for it to happen to.

Just as our emptiest rooms turn out to be dense with invisible radiation (you have only to tune in the appropriate receiver - radio, television - to make it appear), so the light of a billion stars, the background hum from the Big Bang and ever fainter echoes of countless pasts are expanding through one another everywhere, making the hollow night claustrophobic. Even if you speak not in terms of the something of energy but the something of substance and ask: must there not be a place - numberless places - where there is no matter at all? - you will hear a dusty reply from our cosmologists. Empty space itself, theory suggests and measurement indirectly confirms, is a lusty begetter of virtual particles: for mirroring the rivalry in mathematics between continuous and discrete approaches to zero, theories of space seem to oscillate between an energised vacuum and a sea of all but things.

So when the orthodox theory of ether held sway (the theory that explained light as waves in an all-pervasive, rarified fluid), you would have expected - given the pushmi-pullyu of fashion - that a contrary hypothesis would have arisen which filled up space with celestial ball-bearings: and you would have been right. In that corner of the library where no young and hardly any old codgers go are stored the books of yesterday's certainties, the delicate charm of their fustiness ruffled only by the ever modern authority of their tone. Here you will find - almost invisibly wedged away - Osborne Reynolds' slim volume, <u>On an Inversion of Ideas as to the Structure of the Universe</u>.

Its date is 1903 and its writer - an English professor of engineering - assures us that we may have the fullest confidence in his explanations, whose success will achieve "that ideal which, from the time of Thales and Plato, has excited the highest philosophical interest." The reason for his confidence is that his 'inversion of ideas' perfectly explains, he says, the basic phenomena of physics (the absorption and reflection of light, gravitation, etc.), and "Considering that not one of these phenomena had previously received a mechanical explanation, it appears how indefinitely small" (zero à la mode) "must be the probability that there should be another structure for the universe which would satisfy the same evidence."

And what is his explanation? Why, that there is no empty space, but that the universe is all compact of inconceivably small grains (their diameters are the seven hundred thousand millionth part of the wave-length of violet light), moving at about one and one-third feet per second relative to one another, along very short paths. They make up an elastic medium that extends to infinity, and in their piling account for everything we know. Well may Reynolds ask: "Could anything be more simple?" His theory explains, for example, why the sky is black on a clear night: the motion of light is dissipated to increase the relative motions of the grains. It accounts for gravitation as the result of the particles no longer being closely packed, due to inward strains. It even explains what we are: waves in this granular universe; "...that is what the singular surfaces, which we call matter, are - waves. We are all waves."

O Archimedes, this is wondrous strange! Where are your poppy-seeds, and Buddha, your motes that danced in a sunbeam, now? A universe in which the void has disappeared - and not only the void, but Osborne Reynolds too: "I have in my hand," he writes, "the first experimental model universe, a soft indiarubber bag..." And there is a photograph, his Figure 9, of the universe (marked W) gripped in a hand that extends from a white cuff, itself shot from a dark sleeve. But the photograph ends before the elbow. There is no more. It is the only likeness I have of Osborne Reynolds and his universe.

A quarter century later, when quantum mechanics was carrying all before it, another inversion of ideas gave us positively charged particles as holes in a universal, undetectable sea of negativity. This was P. A. M. Dirac's way of saving the fundamental wave equation that bears his name, and which describes how electrons move, from some very nasty consequences (such as electrons growing ever more negative and radiating away infinite amounts of energy as they did so). It was a solution - as one physicist put it - "that made up in brilliance what it lacked in plausibility", and there is certainly something to be said for seeing the positive as a zero in negative contexts. Dirac's sea, however, went the way of Reynolds' shore when the negative energy solutions were later rethought as positive energy states of a different particle.

Now, as we understand it, the deepest freeze in the uttermost basements of space is around 2.7 degrees above absolute zero, signalling the presence there of something in motion. More (or less, since less is more): even a vacuum empty of all matter is still a surging ocean of energy in the form of fluctuating electromagnetic waves, as shown by their exerting a force on uncharged plates (a prediction made in 1948 by the Dutch physicist Hendrick Casimir and finally confirmed in 1996 at Los Alamos). The amount of this "vacuum energy" is (troublingly) infinite, according to current theory; but thus far only the little people in flying saucers seem able to tap it.

If we aren't to find nothingness in the great open spaces, what about the minute closed ones we make at home, under the bell jar? In 1670 Robert Boyle, the English chemist and alchemist - founder with his fellow natural philosophers of a society they called The Invisible College - put a bird in a glass and pumped out the air. You see it fluttering still in the painting Joseph Wright of Derby made a hundred years later: the little girl of the family turns away weeping; her brother pulls up the empty cage on a pulley; the adults stare at one another, at the experiment or into their thoughts - only the flowing-haired figure at the apparatus looks out at us with tense inquiry, and we look rivetted back.

"Is it enough?" he seems to be asking. Yes, as far as killing the bird goes; no, if he wanted a perfect vacuum. Generations of quiet, ingenious workers have brought us much closer. You wouldn't guess, passing them in the street outside their laboratories, that to them the world is well lost for one more zero in the decimal perfecting of their results. They live their work: William Fairbank - long the doyen of near-zero researchers - slept with his windows open to frigid winds and dusted the snow from his bedding in the morning.

These days physicists use lasers to stop rubidium atoms dead in their tracks and so super-cool them in an evacuated glass thermos, then trigger a magnetic trap to cool them further, to within 100 billionths of a degree of absolute zero. At this point those atoms act collectively as one; yet even were their temperature precisely zero they would, paradoxically, have not quite zero energy. This follows from quantum mechanics, at whose core lies Heisenberg's Uncertainty Principle, which says that when it comes to an electron, you can at any moment know where it is or how fast it is moving, but not both - but you would be pardoned for thinking that rather the principles of Mahavira's syad-vada were central here, and this all but empty world of minute appearances was at the same time real, unreal and indescribable.

Following Kant, it is clear that we can imagine space without objects but never objects without space. But modern physics at its macroscopic and microscopic extremes seems to have shown that we cannot think of actual space as empty. Doesn't its bending around to cradle galaxies, its acting as a matrix for particles spurting into and out of existence, accord well with our revised understanding of the Hindu void, sunya, not as empty but pregnant?

If we seek a simpler, purer zero out there, we may have to look for it in a different guise. What about the *mass* of certain particles? Some physicists are sure that speaking of a photon's or a graviton's mass is meaningful, and that this mass is zero - but it isn't anything they can prove.

Turn then to what surely has a zero stamped on it: time, which of all things best deserves to be called the unmoved mover. Yet here again we are likely to be frustrated: what we hear of the noise celebrating that first new year is a Big Bang after a few nanoseconds which, for all our ears and minds, we cannot recapture. Or was it more than a nest of instants - did billions of years or even an infinite stretch of time precede the birth of our universe?

For the cosmological theory we've been soldiering along with over the decades is in crisis: recalcitrant facts are breaking it up into almost as many patchwork hypotheses as the universes some think are jostling one another, mere bubbles in a beery froth. Perhaps it is about to be downed by a despondent medical student named Prior, as John Collier's story, "The Devil, George and Rosie", explains. Don't worry, the Devil says: "...it will be twenty million billion years before his lips reach the glass, for a young woman is fixing him with her eye, and by the time he drinks all the bubbles will be gone..."

Some cosmologists give time no beginning, some would have it perceived differently within and outside a bubble universe, some see our particular bubble's time as linear within the larger universe's cyclic time, and yet others picture it "before" the Big Bang as behaving like space, with no headlong direction. These scenarios follow from the varied ways of rejigging a derelict theory to fit new observations. They make you marvel once again at the power of language and abstraction, since time seems so much a condition of all our doing that you wonder where you must stand to see it thus from afar. Time, to paraphrase Kant, is something we think with, hence only perilously about.

Could we find zero perhaps in a cogent answer to the related question: what did the world begin *with*? Here too there are numerous knockers at the gate. One physicist says: "The reason that there is Something rather than Nothing is that Nothing is unstable."

St. Augustine wrestled hard and long with this problem in his <u>Confessions</u>, all too aware that our questions outpace our answers and that many different senses can and ought to be extracted from a text. He concluded that the earth and the deep, which in the beginning were invisible and lightless and without form, God must have made from almost nothing, *prope nihil:* "...not altogether nothing; for there was a certain formlessness, without any beauty." A wonderfully subtle position to take, by a man whose thought has so informed ours. Should we therefore be surprised to find how similar the view is sixteen centuries later? As the universe cooled after the Big Bang, matter and anti-matter almost annihilated one another, to leave pure radiation. The symmetry wasn't perfect, however (it was without any beauty): for every 100 million pairs of quarks and anti-quarks there was one extra quark, the foundation of matter - and this faintest tip of the balance evolved to stars and planets and seagulls and ourselves.

The ends of space and the beginnings of time: it turns out zero wasn't crouching in either of these trees we have been barking up. More natural, surely, to look for it out there at the dead center of things. That was how zero came to be understood as the negative numbers grew increasingly vivid: the fulcrum they and the positives balanced around; and, more broadly, as the linchpin of coordinate systems that let us navigate safely out and back again.

Where is this center? Zeus found it by starting two eagles at the same moment from the eastern and western edges of the world: they met at Delphi, and the sacred stone there marks the world's omphalos. Might that stone - or navels in general - have contributed to the form of zero, accounting perhaps in particular for the peculiar we met in medieval times? Long after Zeus, Aristotle took the Pythagoreans to task for putting the sun at the center of the universe on purely theoretical grounds (fire, being superior to earth, they said, deserved the honor). His observations led him to affirm what most knew, that the earth was at the center - where it remained for many centuries.

The center of this center was called by early Muslim astronomers "The cupola of the earth" - by the Hindus, the island of Lanka, their 0° longitude and (charmingly) without latitude. It was there that the demon Râvana built the labyrinthine fortress Yâvana-koti, whose plan is strongly reminiscent of the palace made by the sons of Atlas in Plato's Atlantis. But Atlantis sank beneath the ocean for its

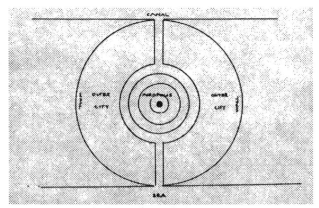

The plan of Atlantis

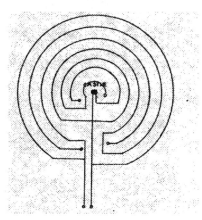

Yavana-Koti

sins, and there is no such cupola, said the traveller Albiruni a millennium ago, while Lanka he guessed to be the cannibal island Langa, source of cloves and the wind that brought small-pox.

Even the notion of the earth as midmost of all came under a series of diverse and remarkable attacks before Copernicus dealt it a lethal blow in 1560. The Stoics, for example, in the generations following Aristotle, saw the material universe afloat in an infinite void. The former might have a center but the whole clearly could not. Long after, near the end of the middle ages, the Cardinal of Brixen, Nicolaus Cusanus, claimed that the physical universe was by its very nature imprecise, so that neither the earth nor the world itself could be a true sphere since a more perfect one was always possible. Being indefinite, therefore, none had a center. In fact the true center of the universe, he said, coincides with its circumference and neither is physical: that perfection belongs alone to God. Because no mortal can be at this center, all of our differing views are equivalent - and equivalently lacking in objectivity:

Combine thus these diverse imaginations...and then, with the intellect, which alone can practice *learned ignorance,* you will see that the world and its motion cannot be represented by a figure, because it will appear almost as a wheel within a wheel, and a sphere within a sphere, having nowhere, as we have seen, either a center or a circumference.

If you want to take in at a glance how the center of western thought shifted over five turbulent centuries, just compare three treatments of this metaphor, with that of Cusanus acting as their mean. The first is from the 12th century <u>Book</u> <u>of</u> <u>the</u> <u>XXIV</u> <u>Philosophers</u>, where God is described as a

Sphaera cuius centrum ubique, circumferentia nullibi:
a sphere whose center is everywhere, its circumference
nowhere.

The passage quoted before from Cusanus, written in
1440, continues:

> It is as if the machine of this world had its center
> everywhere and its circumference nowhere, because
> the circumference and the center are God, who is
> everywhere and nowhere.

By 1660 only the machine is left. Pascal writes in his
<u>Pensées</u>:

> The whole visible world... is an infinite sphere,
> the center of which is everywhere, the circum-
> ference nowhere.

So we pass from a finite, through an indefinite, to an
infinite universe.

With Copernicus the center contracted to the sun, and
there it still was fifty years later for Kepler, and for Newton
later still. Between them, Galileo was cautiously sceptical:
"We do not know where to find the center of the universe
or whether it exists at all." Since it was politically incorrect
to express such doubts (and metaphysical scandals then
carried far heavier penalties than physical scandals now) -
and perhaps because Galileo saw no way to test them - he
left the question alone.

Not so Leibniz, who carried his attack on Newton into
space. For Newton, space was the shining frame which
made sense of the relations among bodies within it; for
Leibniz, space was just the order of these related bodies
and wouldn't exist without them. Unlike the two women
shouting from their Edinburgh windows at one another,

who would never agree (says Sydney Smith) because they were arguing from different premises, Newton and Leibniz would never agree because their premises were the same: God needed the greatest glorification. Absolute, empty space, Newton thought, implied God's continual governance of all within it; this read, to Leibniz, as God always having to wind up a faltering clock.

The arguments themselves wound up and down, changing terms, tropes and proponents - then were swept away at a stroke two hundred years later by Einstein. For Leibniz saw that Newton had conjured up absolute space in order to talk (as in his First Law) about absolute motion in it; but that no such motion could ever be observed, since you would detect no change - hence the fiction of absolute space was unnecessary. Einstein stood this remark on its head: what one observed *from,* he realized, are private frames of reference, each bearing its own center, its zero, about with it. As long as the frames were unmoving, or moved uniformly with respect to one another, each would appear at rest to its occupant, and while events seen from them would be measured as occurring at different times and places to different observers, the laws governing the events - the laws of physics - would read as indistinguishably the same. This was Einstein's Principle of Relativity, which thus neither affirmed nor denied the notion of an absolute space with a fixed center, but made it meaningless.

Has this parade of just a few rival cosmologies made you doubt cosmology altogether, or suspect it to be an annex of theology? Do you worry that this science harbors its own Uncertainty Principle which lets us determine either what a position is or how fast it is changing? The difficulty lies in trying to sketch out the universe by connecting too few datum-points. Does the last elephant holding the universe up stand on a turtle, or is it elephants all the way down? For despite the confidence of Osborne Reynolds, any number of interpretations will be consistent with a limited body of evidence. Which theories arise and thrive may involve something more than how well they fit their curves to the facts. If the temper of the times is aesthetic, elegance outranks sheer plod, and brilliance more than makes up for plausibility. Even the leisure pursuits of a critical community may incline its science toward hypotheses that resonate with science fiction.

More fundamentally, when at the very edge of your territory (and it is always on the edge that science moves) you may believe as settlers do that the world afar is like the world at hand, and so strive for analytic continuations of known laws into that unknown; or like an explorer you may seek to distance yourself from a provincial past, rejoicing in the essentially singular. Nothing new under the sun? Then beyond it! And while we are all over-reachers on a negligible speck of dust, the cosmologists among us seem to belong to the second camp.

Yet settler or explorer, they are scientists in the end rather than theologians - and rather than literary critics. They read the book of nature without benefit of old or new criticism, semiotics or structuralism, believing what as humans they can't help but believe, that what is out there is unequivocal and - unlike scripture - independent of our interpretations. Their ends may be hubristic but their means are modest, and in their mind's eye they still wear the white lab coats of the objective, detached observer. Not for them the truth of formalised mathematics that there are essentially different models of the same phenomena: they resemble the mathematicians inventing rather than validating relations, and recognizing with a shock that these inventions are discoveries within the singular way things are.

Why, then, is cosmology's evolution punctuated rather than continuous? Because there is - as William Whewell said some hundred and fifty years ago - a mask of theory over the whole face of nature. It is this mask scientists look at, rather than through: a mask of papier-mâché, pasted together from fragmentary facts that compile to features so far from those they first crudely conformed to. It is a mask whose layers crack and peel apart when its flour-and-water reasoning dries out.

Yet zeros like eye-holes pierce this mask in many places. We saw an instance before where they signalled the maxima and minima of processes. The deepest laws of physics are written in what Napier, you remember, called "equations to nothing": those laws of conservation, I mean, that say the total energy (or charge, or momentum or matter) in a system remains the same: its changes amount to exchanges that sum to zero. These laws go back at least as far as Descartes, who held that what God created He then preserved. Although the objects that physics studies variously altered or grew more precise and even multiplied over time, the laws expressing their conservation were themselves conserved. These laws and their like aren't decorative incisings on the pillars holding up our understanding of the world: they are the stuff of the least brick, alone letting us build a detailed image in the likeness of things - half of whose texture is therefore these zeros.

A telling example: it may not be all that easy to study the forces poised in bodies (such as a system of pulleys) at rest - in equilibrium - but it is a great deal easier than analysing those of bodies in motion (swinging, sliding, rotating, rebounding, falling). Any way of reducing such dynamics to the former statics would clearly be welcomed. This is just what Jean Le Rond d'Alembert did in 1743 - and did by a change so slight in the mere *form* of an equation that we relish anew the power of grammar to yield insight (which, like faith, follows from just going on tinkering).

For Newton's Second Law of *Motion* asserts that moving force is the same as the mass of the body moved times its acceleration:

$F = ma$.

This puts the three terms in a mutually defining dance that promises severe difficulties in analysing such forces. What d'Alembert did in effect was just to rewrite Newton's equation as

$F - ma = 0$,

then to consider the term "$-ma$" as itself a force, "the force of inertia":

$I = -ma$. So now we have:

$F + I = 0$.

But this states that the sum of the forces, F and I, produces equilibrium: and hence dynamics is reduced to statics. Of course equations with specific forces and "forces of inertia" still have to be solved - but they can now be *understood* in the same way and from the same vantage point that familiar problems of equilibrium were.

Thus d'Alembert's Principle, that any system of forces is in equilibrium if we add together their applied and inertial varieties, has the character we recognised long since in double-entry bookkeeping: a balance is brought to zero with the aid of a little artifice. Magical helpers have magical helpers to help them.

You will rightly say of these conservation laws, as of d'Alembert's Principle, that being relations among things rather than things themselves they aren't truly out there, hence neither are the zeros they involve. The virtual particles we began this chapter with are virtual not only in being taken almost as particles, but in being almost *taken:* for physicists take them more as mathematical metaphors, aids to conception and computation, very much like the infinitesimals of calculus that help us determine the slope of a curve at a point on it, but disappear once the calculation is done. Shall we temper Leibniz' claim that only individuals are real and relations belong fully to thought? Let's say instead that zero is a relation and relations lie in that middle ground between mind and matter where mathematics strikes its accords, making interpretation and reality available to each other. Surely that is why these smaller-than-life figures of zero are always busy around the fringes of our affairs.

There they are on our thermometers, for example: but Fahrenheit was a man before he was a scale, choosing his zero as the lowest temperature attainable by a freezing mixture of ice and salt. René de Réaumur - when he wasn't studying the forms of birds' nests or proving that the strength of a rope was less than the combined strengths of its strands - argued for making zero the freezing point just of water. Anders Celsius observed the aurora in Lapland - well situated to devise the Centigrade zero; and William Thomson, Lord Kelvin, a blur of ceaseless calibrating the whole of his eighty-three years, set zero degrees as the temperature at which motion ceases.

Man is the measurer of all things. Look at the Plimsoll Line, there on ships the world over to save the lives of sailors by showing the level between safe and dangerous loading: Samuel Plimsoll came near to capsizing his own fortunes when he shook his fist in Disraeli's face and called his fellow members of Parliament villains for dropping the bill that would have outlawed the infamous "coffin ships". Look at our midnights, our meridians, look at what scales you will, they are all made with invented boundaries we almost believe are out there, but mark instead the personality of their inventors.

Still, might there not be a membrane of outside patterned with pores of zero - not as remote as the cosmos yet sufficiently far away to count as Out There? For we fall from time to time into Cartesian reveries where our bodies strike us as the bearers of what we are: inertial reference frames for the passenger mind. Professional athletes often talk of their accomplishments in this odd, third-personal way: evaluations of what the arms, the legs were up to, of a kind with racquet, ball, glove.

We often fail to savor to the full the astonishing virtuosity athletes show every day in diving for a ball and coming up throwing; in stretching for an overhead smash and landing ready for the return; in passing their centers of gravity smoothly under the bar they vault themselves over, or lithely leaning into and out of an ice-slick curve. Cat-like, they are geniuses at always having their bearings: a sensation we ourselves grow aware of in dancing, say, or cycling, and know without knowing through the thousand gestures of the day. Zero in the body as machine shows as its center. You have only to recall those sickening moments of vertigo, or read about people suffering from inner ear imbalances or the catastrophic loss of their sense of position, to realise how casually, how vitally, the gyroscope of zero holds each of our voyaging reference-frames on course.

Is there, like the dot within the circle of one of the zeros we came across (that theca or theta), a different zero in the mind from the zero of the body that carries it about? Nothing like a balance - closer, perhaps, to the center of Pascal's infinite sphere, which is everywhere: since each of us correctly knows that ours is the unique center. We call it 'I', and suppose that others, in some way that imagination alone can grasp, mean by it for their selves what we mean by it for ours. Recognising that they say 'I' in the same way we do severs at the same time that it connects us definitively to the world.

It is a paradox we accomodate ourselves to without ever resolving. So a singer brought up to practice the scales with C always taken as 'doh' is disconcerted to learn that in another tradition 'doh' names the first note of whatever key a piece is in. To protest that nevertheless 'doh' is really C and that 'I' rightly belongs only to me is to run foul of of the movable 'doh' and the indiscriminate use of the personal pronoun. We may never manage to do more than translate these alien usages into our proper, solipsistic, tongue.

And this zero of self - since it lies behind sense, do we grasp it as absence? If it is the grasper, do we grasp it at all? Perhaps in stalled moments; perhaps as a flicker out the corner of the eye; perhaps as a shadow in the slanting light of fear, never coming on a snake

> Without a tighter breathing
> And zero at the bone.

Danger makes us know that we store something precious - but not what it is. We bear a message we cannot read, like those Greek runners who carried the long thin ribbons that had been written on when they were wound around a staff, and whose words would only make sense when the letter halves came together again on the matching staff of the receiver.

13

BATH-HOUSE
WITH SPIDERS

Zero is neither negative nor positive, but the narrowest of
no-man's lands between those two kingdoms. Yet our anal-
ogy-driven minds, ever eager to read expressions on neutral
faces, seize on its emptiness and make out powers and por-
tents there. Let's ask first what expressions they find when
expecting the worst - then in the next chapter how this
same emptiness can appear benign.

You saw how, in the middle ages, zero was taken as
devil's work or the devil himself, the great canceller of
meaning. How easy to think of yourself dismissed as a
good-for-nothing, a no-account. The Sultan Abdul Hamid
the Second, perpetrator of the terrible 19th century Armen-
ian massacres, had his censors, they say, remove any reference to H_2O from chemistry books entering his empire,
convinced that the symbol stood for "Hamid the Second is
Nothing!" How easy still to dismiss those we reject as mere
zeros, sinks of energy, black holes in which all that matters,
the singular and the memory of the singular, disappear
without a trace.

It is as if that hollow oval stood for anonymnity, mirroring our fear of making no difference to others - to anyone -
to the world: "...to pass beyond and leave no lasting trace",
wrote William McFee in <u>Casuals of the Sea</u> (which left little

enough trace itself: the colors that shimmered in the bev-
elled glass of the school's specimen case).

How many zeros it turns out you've met and even kid-
ded around with: those faces from old yearbooks you now
can't put a name to, those names in old address-books that
summon no features. They made up the crowd, into which
sometimes you yourself loved to plunge as a mere spectator
- until the chilling thought came of your own name as
unrecognisable in tattered address-books lying in basement
drawers.

To live as a zero: the superfluous man, the man without
qualities, the person who, like Henry James' John Marcher,
finds out only too late that the beast in his jungle was that
there was no beast, no response to passion: this figure
haunts our fiction and our fact, the *salaryman* of Japanese
society, the company man, the fungible folk of our office
culture, who retreat to virtual reality games at home. The
worst of it is that, unlike Marcher, most never awake to the
truth that they haven't lived.

This was *le néant,* the nothingness at the heart of Existentialism: the rare recognition of which brought nausea -
and so a moment in which to choose what you would be
(since all choices, being choices, were equally unnecessi-
tated, equally arbitrary, equally without prior meaning).
You began to exist with this random choice, embraced as
random, and in living it out you made your essence: Sartre's
neat reversal of the Thomistic formula that essence pre-
cedes existence. So an arbitrary digit slipped in front of a
string of zeroes creates value where none had been. But
Sartre died and his authenticity died soon after, as pipe and
scarf and the Deux Magots came to seem the self-publicis-
ing tactics of a literary man. The fashion for Existentialism
passed, preserved now only in successive waves of adoles-
cence, before the getting and the spending set in.

At least those who haven't awoken are lost in their dreams. There is a far more fearsome embodiment of zero: the guilt-ridden certainty of one's own utter worthlessness. The self-damned knew even in childhood the etymology of their naughtiness: it derived from being nought. The least whispered nuance of scorn singled them out; they heard themselves as Hamlet, asking what they should do, crawling between earth and heaven. When Donne addressed man as that Nothing, infinitely less than a mathematicall point, than an imaginary Atome, it was to each of them he spoke, cut off on their islands. But what were pulpit thunders to the silent certainties, in the dark night of the soul, that zero was only the beginning of endless negation?

Over this nether millstone grinds another in contrary motion: detest your self as you may, how devastating to think of losing it. By an act of mind no logic can sanction yet fantasy distills, you see the world as it would be without your presence to see it, and despair as John Bunyan did:

> ...I was more loathsome in my own eyes than
> was a toad... I was both a burden and a terror
> to myself; nor did I ever so know, as now,
> what it was to be weary of my life, and yet
> afraid to die.

How could your name possibly be writ in water, as Keats, dying, said of his? How could its syllables ever weaken away in the on-going hum of things? A book on the latest cyberspace cryptography tells us that it

> is about submerging a conversation in a flow
> of noise so that no one can know if a conver-
> sation exists at all. It is about taking your
> being [and] dissolving it into nothingness...

Of course the author continues:

...and then pulling it out of the nothingness
so it can live again.

How can we be certain this promise will be kept? Have
you never lost for good a message on your e-mail? Animula
vagula blandula, the little soul wavers away, wrote the
Emperor Hadrian. How irretrievably unjust that yours,
stored with its singular cargo, should break up forever. Or
is it the puzzling ship of Theseus that every sleep pulls apart
to its least peg and every waking rebuilds with new matter
in the old form, so that there is no I to your I, no continuing self, but successive semblances that fade and wear out
at last?

These circles of negative nothingness may hang for years
above us - yet as Sylvia Plath wrote:

How did I know that someday - at college,
in Europe, somewhere, anywhere - the bell
jar, with its stifling distortions, wouldn't
descend again?

Could there be any zero more negative than this? One:
the world and everything in it (hence, trivially, yourself too)
picked bare of meaning. It takes no more than looking
down at the furrow you followed and discovering it is the
track of a million-year glacier; it takes just drawing aside
the blue veil of sky and seeing the mindless expanse of
cause and effect: suddenly all that shone now only glitters.
When the philosophy building at Harvard was near completion, some of its faculty wanted the motto over the
entrance to read: "Man is the Measure of All Things." At
the unveiling, however, they found carved in stone: "What
is Man that Thou art mindful of him?"

it isn't so much a disclosing by nature as a closing off of inner life that leaves the world monochrome. You withdraw (defensively, perhaps) your interest from yourself - for a time, you think; or from someone you now once loved. The chill, unbidden, widens to your kith and then your kind. The universe looks lovelessly back at you. You display a while your worldly disenchantment; then a sophisticated cynicism; then cynicism plain; then disengagement leaves you in your corner, visited now and again by grotesque visions: "as soon as one is ill," says Svidrigaïlov to Raskolnikov in Dostoyevsky's <u>Crime</u> <u>and</u> <u>Punishment</u>,

> as soon as the normal earthly order of the organism is broken, one begins to realise the possibility of another world... And what if there are only spiders there, or something of that sort... We always imagine eternity as something beyond our conception, something vast, vast! But why must it be vast? Instead of all that, what if it's one little room, like a bathhouse in the country, black and grimy and spiders in every corner, and that's all eternity is?

Do you find this claustrophobia worse, I wonder, than agoraphobic boundless emptiness? The mathematician and physicist Hermann Weyl suggested that when the ego is extinguished, the unmarked grid of coordinate space remained - the infinite, centerless space you saw in the previous chapter, which we in our moving inertial reference frames make arbitrary pools in. A devastating picture is in Ford Madox Ford's novel, <u>The</u> <u>Good</u> <u>Soldier</u>:

> ...upon an immense plain, suspended in mid-air, I seem to see three figures, two of them clasped in an intense embrace, and one intolerably solitary. It is in black and white, my picture of that

judgment, an etching perhaps; only I cannot tell
an etching from a photographic reproduction. And
the immense plain is the hand of God, stretching
out for miles and miles, with great spaces above
it and below it.

No matter how few or many, how ill or well conceived,
distinctions everywhere and anywhere have made for
meaning. It is only when they rupture, leaving a background with nothing on it or figures against no ground,
that negation floods in. Meaning needs a content set in a
context, which needs in turn what holds the two apart. It is
as if in these latest excursions we had mistaken the hollow
within its ring for zero, or took zero to be the surrounding
space that the ring shut out. But zero is neither - it is the
ring itself: pure holding apart.

14

A LAND WHERE
IT WAS ALWAYS
AFTERNOON

Those for whom all this nothingness is oppressively real take comfort at least in this, that extinction means the end of oppressive reality. They read Schopenhauer in their overcast Novembers with a kind of grim satisfaction, and return the knowing smile they see on the faces of eastern gods: all is vanity. Zero becomes for them at least not negative. A friend told me of a passage in Beckett where one of his characters, toting up the puny sum of things, concludes that at least it's better than nothing.

"Really?" says the other, surprised, "Better than nothing? How can that be?"

This hankering after nothingness as a relief always rings somehow false, since it paradoxically supposes your presence enjoying your absence. Such an ill-imagined scenario lies behind many misinterpretations of Nirvana as the Bliss of Non-Being, and other upper-case abstractions. It accounts in part for the hollow sonorities of Swinburne, who thanks with brief thanksgiving whatever gods may be that no life lives forever, and whose weariest river winds somewhere safe to sea. But there is here too a sensuous delight in syllables rather than ashes on the tongue, and perhaps the hint of a different pleasure in vestal disguise.

Don't you feel like saying, with Melville, to all this sallow tribe, "Give it up"? Yet as he adds, one sometimes loves to sit with these poor devils of sub-sub-librarians "and feel poor devilish, too; and grow convivial upon tears."

Even more delightful - as we watch the zero of the heart slowly changing its sign from minus to plus - is just to sit, just to veg out: that wonderfully expressive tribute the seventies paid to the mindless life of plants. Yet it isn't exactly mindlessness we crave - not the declension from hebephrenia to catatonia - rather, having our wits about us but using none. What a bizarre ideal, an extra-terrestrial visitor might think - but we know it in ourselves: dozy days, days at the beach, days curled up with a jug of wine, a loaf of bread and a paperback romance, one with the lotos-eaters in a land where it was always afternoon.

How sweet it were, hearing the downward stream,
With half-shut eyes ever to seem
Falling asleep in a half-dream!

Why should a pathology so counter to our hectic plungings be sweet? Old explanations are sometimes still the best: conspicuous leisure affirms our superiority (even if the audience is only ourselves). Or to move the motive from the economic to the psychological: potential power always seems greater than actual - perhaps because it is measured on a scale that stretches as easily as itself.

These nothings so far have registered barely positive because the states of mind they expressed were all passive. Begin to animate the picture and you will see zero accumulate charge. Certainly ceremonies of atonement restore to zero the ethical bookkeeping of the year, when everyone counts as gain the erasing of cankerous debts. How striking that in a world run on retribution, good can be made again by no more than forgiving bad.

A very different sort of gain is sensed by many to lie in reducing themselves to zero, humbling their pride, abasing their bodies like those emaciated saints of the middle ages. But even when such observances begin with calculation, ritual develops a momentum of its own: a kind of abstract sensuality lifts the weightless spirit up and the forms of devotion grow more real than the matter that clogged them.

Some, however, incline by temperament to humility without a thought for being first by being last. Shyness, gentility, the ecstasies of subordination commingle here. The pitiful simple heart that Flaubert described in such pitiless detail beats in a million rooms stuffed with souvenirs of a life lived for others. And if you look up "Nothingness Itself" you will be cross-referenced to the Venerable Father Antonio Margil de Jesús, who so styled himself: La Misma Nada. He was a Franciscan missionary in the American southwest three centuries ago, convinced that anything less than total abnegation would rob God of his glory. He called Mary La Doña Nada, Lady Nothingness. The indians he converted in Central America were descended from those to whom zero was the god of death.

It says something about the fine-tuning of our moral antennae that we are so apt at telling the Father Margils from those for whom humility is a strategy for salvation. We always catch beneath a spoken "not as holy as Him" the silent sub-text: "...but holier than thou." Two rich men, the story goes, were outdoing each other in the Temple, protesting their insignificance: "Oh Lord!" said one, "I am less than the spider-web strung in the light of this sunbeam, compared to You!" "But I am less even than the tiny spider that spun it!" declared the other. Just then a poor man entered and was dazzled by the sight of the almost transparent, glistening strands. "Lord!" he cried out in rapture, "how glorious are Your works! Why, compared to You, I am less than the grains of dust that catch in this web!" One of the rich men nudged the other and whispered: "So look who's claiming he's nothing!"

The ascent to sanctity is hard, but it seems somehow unfitting that so much effort should have to go into reaching zero, for all that it shines like a halo. Either you stumble on the stillness of things, or you walk toward it on a leveller path as in Taoism, for example, or Yoga: calming the inner monkey, damping down the waves of fervor and despair to the steady state in which you can hear again nature naturing. Figure and ground reverse: not zero but all the noise becomes non-being, the cotton wool Virginia Woolf so vividly described:

> Every day includes much more non-being than being.... the goodness... embedded in a kind of nondescript cotton wool... One walks, eats, sees things, deals with what has to be done; the broken vacuum cleaner; ordering dinner; writing orders to Mabel... As a child then, my days, just as they do now, contained a large proportion of cotton wool... Week after week passed at St Ives and nothing made any dint upon me. Then, for no reason that I know about, there was a sudden violent shock... I was looking at the flower bed by the front door; "That is the whole", I said. I was looking at a plant with a spread of leaves; and it seemed suddenly plain that the flower itself was a part of the earth; that a ring enclosed what was the flower; and that was the real flower; part earth; part flower.

Ridding life of its cotton wool - or more urgently, clearing away some of the clutter in an irremediably dirty world: zero's value increases as purification recurrently gleams on our horizons. It takes various forms. Some wash their faces to wash away their sins, since inside every luxuriating soul is an ascetic soul trying to get out. Others wash the face the world shows them, like the obsessive-compulsive patients psychoanalysts hire to clean their houses.

For some, however, the religious motive becomes aesthetic. Here the insight that less is more opens at the center of understated art. The last tail-feathers of migrating cranes composing the white expanse of a Japanese screen; the Scandinavian ideal of bleached wood and snow; brief wit; pithy asides. The minimalist arts are presided over by the unadorned Graces of simplicity, innuendo and elegance, who steer philosophy too, urging apprentice thinkers to empty their minds that the truth may better be inscribed there. They charm those mathematicians who delight in reducing the gigantic sprawl of theorems down to ever more spare foundations, a set of laconic axioms at last; as well as those for whom it is all a play of pure forms drained of meaning. Yet people to whom meaning matters and who recognise that content and context make each other meaningful, know that when each is pared to a minimum, a dab of one intensifies the other - and a context of nothing at all pins everything on the minutest presence.

An empty presence, turn for turn, must bring the context into focus: and this is the zero ideal of those who would be invisible. Their motives differ: hiding out amid the banditry of life; exerting a real or imagined control from under the cap of darkness or behind the one-way glass; the studied banality of the spy, the reporter's casual anonymnity. Most complex of all, perhaps, is the writer's sensibility: a kind of huge spider web, Henry James called it, unseen but catching everything that goes past and converting the air's very pulses into revelations.

Do such attempts at transparency seem too contrived, too labored at, rightly to count as a blissful gliding into being no one? They sound more like the fierce rivalry of the friends in Jack London's unmetaphorical story "The Shadow and the Flash", who try to outdo each other in making themselves invisible - one by absorbing, the other by reflecting light (their presence mortally betrayed, in the end, to each other, as the title foretold). Prepare as you may for your personality to become invisible, its actually doing so is always unexpected. Emerson was crossing the common one winter twilight:

> Standing on the bare ground, - my head bathed by
> the blithe air and uplifted into infinite space, -
> all mean egotism vanishes. I become a transparent
> eyeball; I am nothing; I see all; the currents of the
> Universal Being circulate through me; I am part or
> particle of God. The name of the nearest friend
> sounds then foreign and accidental: to be brothers,
> to be acquaintances, - master or servant, is then
> a trifle and a disturbance.

in reading this are all those 'I's are a bit disturbing: how selfless can so reiterated a self have been? It is curious that each of us regularly makes up an impersonal persona, a third person 'one' to witness objectively from. And we value - to the point of intensely personal invective - markers that testify to the sexlessness of office, to the ungendered bearer of traits. Alternating 'he's and 'she's in a text, Aristophanic s/hes and s/hims reminiscent of nothing so much as Barthes' s/z, 'person' for the superordinate 'man' and 'e' or 'ha' offered as an androgyne: should we just say 'Dorothy Parker' and be done with it? For she described a party she had been to where seven sexes were present: male, female, homosexual, lesbian, hermaphroditic, neuter - and herself. The real loser in all this wrangle is clearly zero, which - had it a voice - would protest much more loudly than any at a hollowed-out self being commonly called 'one'.

Yet what are these scramblings and slidings and glidings toward zero compared to the effortless pleasure we daily engage in, accomplished by thousands of years of evolving culture recapitulated in our separate childhoods? I mean the pleasure of reading, where he who loses himself shall find himself as another and another and another, or as something more than an angel hovering over the scene. This elevation is what the invisible writer gives to the anonymous reader, without whom all the disappearing acts were in vain. At dinner once Henry James answered the murmured questions his admiring neighbor put to him about his novels, then turned to her in amazement: "Can it be - it must be," he said, "that you are the embodiment of the incorporeal, that elusive and ineluctable being to whom through the generations novelists have so unavailingly made invocation; in short, the *Gentle Reader*? I have often wondered in what guise you would appear..."

We have gone from valleys to peaks of nothingness, from despair to exhilaration, as zero changed its emotional sign. But could zero ever have been thought of as infinitely valuable, not the nothing out of which God made the world but Godhead itself? Everything can be found in the furrowed landscape of thought, and where better look for so remote an idea than the equally folded Alps? There in the mid 1800s Lorenz Oken lived out the last of his years, exiled in disgrace from his native Germany. He had deduced the whole of physiology, zoology, biology, psychology and geology by pure reason from a few first principles, and I hope these stood by him in Zürich when his friends, his mentors and his country turned away.

I see him swathed in scarves, stamping along the Bahnhofstrasse, a faint exhalation of vapor and thought above him. Here it condenses to the words: "Zero is the primary and eternal act, endlessly positing itself." He pauses by - what is this? Ah, a fellow human, selling würstli. "So God is zero, and zero is infinite intensity." A twitch, a fleeting frown - he walks slowly on, then at a stone lion suddenly stops: "But Man is the whole of arithmetic, the whole of mathematics! Therefore life..." He hesitates, moves unsee-ingly forward - "life..." - there at the Credit Swisse he grasps life: "Life is only a mathematical problem - it ascends ever higher, to Man at last! God is infinite intensity, but Man is infinite extensity! Everything has been created out of the sea-mucus, for Love arose from the foam. Ever more negative numbers, downward through the slime. But upward, ever more positive, past zero to man!"

He turns around, and his thought turns around too: "How could it be..." Up a side street, climbing above the mercilessly clean city: "Because Man is God wholly mani-fested! Man is God conscious of Himself!" The path is too steep for us to follow. We see him dwindling upward, we hear a faint echo: "God = + 0 -, Man = + ∞ 0 - ∞ ..."

There is one last variation on positive zero, and it is in its way even more peculiar than Oken's: for this zero lies always at the beginning of on-going time. It is the Ameri-can Zero. You catch a glimpse of it in our road novels: "Oh look," says Lolita at the moment of metamorphosis, beside her Humbert in their cocoon of a car, "all the nines are changing into the next thousand." You hear it from people who still wake themselves with the thought: Today is the first day of the rest of my life. It is the zero of a society defined by its frontier: and "As successive terminal moraines result from successive glaciations, so each frontier leaves its traces behind it." Frederick Jackson Turner wrote that in 1893. The traces he lists include coarseness and strength, exuberance, inventiveness, selfishness and individ-ualism, an excessive love of liberty and a deficient love of education. Certainly we never learn from history because each of us knows, like the hero of Thomas Wolfe's <u>Look Homeward, Angel</u>, that our chosen incandescence "was borne in upon the very spearhead of history."

Turner lamented the closing of the frontier a hundred years ago - but it never closed. It isn't just that frontiers have opened up since in space or society or technology, but that we all live on the moving margin. We stand like Jefferson at our Palladian windows looking out at the wilderness. Cyclic time is foreign to us, linear time with a past as long as a future riles our cussedness.

"America is the land of zero," said the philosopher Joseph Needleman in Ken Burns' television program on the Shakers: "Start from zero, we start from nothing. That is the idea of America. We start only from our own reason, our own longing, our own search." And Rimbaud - for all that he was French - added the capstone: "Always arriving, you will go everywhere."

15

WAS LEAR RIGHT?

We have come to know zero intimately in its mathematical, physical and psychological embodiments. It remains elusive. Reach down farther toward its roots in the logic that underlies all these. Since so much effort went into the making and rearing of zero and so much effort has since been saved by its presence, can we grasp zero, zero alone, and ask: could it by itself create anything? "Nothing will come of nothing", said Shakespeare's King Lear to his daughter Cordelia, when she refused to join in her sisters' scheming flattery of their father - yet in fact the rest of the play unfolds from her nothing.

Certainly when 0 is coupled with 1 we get the entire world of whole numbers. All our calculators, computers, telephones and televisions - every piece of electronic equipment - operates on the basis of numbers rewritten in a binary code of 0 and 1, corresponding to off and on. This encoding, first hit on by Napier's far-darting mind in 1616, is simple: instead of our ten different symbols, restrict yourself just to 0 and 1, and then bring the full weight of the place-holder system to bear. So after 0 and 1 the next digit - our 2 - would be 10 (not "ten", since we're speaking binary instead of decimal: call it 'one-zero'); 3 is 11, 4 is 100 and so on:

Decimal notation	Binary notation
0	0
1	1
2	10
3	11
4	100
5	101
6	110
7	111
8	1000
9	1001
10	1010
	etc.

You could look at it this way: the Sumerians built on powers of 60, the Mayans on powers of 20 (more or less), we on powers of 10; but the binary system builds on powers of 2. Thus 17, for example, is 16 + 1, which is 10001: that is, one 2^4, no 2^3, no 2^2, no 2^1 and one 2^0:

$$
\begin{array}{ccccc}
1 & 0 & 0 & 0 & 1 \\
\uparrow & \uparrow & \uparrow & \uparrow & \uparrow \\
\end{array}
$$

place: $2^4 \quad 2^3 \quad 2^2 \quad 2^1 \quad 2^0$

If you think of 1 as standing for the whole of things, then what you've just seen is the world of natural numbers generated by the combinings of all and nothing: a metaphysician's dream come true. You can go on to get the negatives, fractions and all the reals from just 0 and 1: for example, -13 is -1101, and 1/4 is .01 (no halves and one fourth). But here is a totally different and very deep game that 0 and 1 can play to yield all the positive fractions as well as all the positive integers. Its legitimate eccentricity shows how many rooms there are in the mansion of mathematics, and how free we are to play in them. "Mathematics *is* freedom", said Georg Cantor, the mathematician who, in the 19th century, opened the doors of the infinite garden to us.

The eccentricity begins with a radical move. Our playing field will be a line on which all these positive numbers will appear, one by one. Set up goalposts just off the field of play, past its two ends. The lefthand goalpost will be 0/1: a perfectly intelligible sign, if you took it to stand for 0, out there to the left, since 0 comes before the positive numbers. Yet don't take 0/1 to stand for 0: don't take it to stand for anything but just to stand, as a goalpost does: a symbol whose two digits we will use in our game played by peculiar rules.

My words of caution are meant to prepare you for the righthand goalpost, which is 1/0. We've seen again and again in our excursions that you can't divide by zero and we're not about to start doing so here. Mathematics is freedom, and for the game we'll play we choose - we *must* choose, as you'll soon see - to set up this hollow expression to the right of our line.

And now we begin to generate all the positive whole numbers and fractions from the two extremes of 0/1 and 1/0. The rule of play is simple: get the first actual number by adding the two numerators for its numerator, the two denominators for its denominator. Since 0 + 1 = 1, our first-born is 1/1, and we'll put it squarely in the middle of the field:.

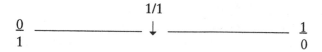

The same rule gives us the two numbers of the next generation: always adding the numerators of neighboring terms for the new numerator, their denominators for the next denominator - and moving from left to right - we get 1/2. in the middle of the left half (since

$$\underline{\frac{0+1}{1+0}} = \underline{\frac{1}{1}}) \text{ and } 2/1 \text{ (i.e., } \underline{\frac{1+1}{1}} = \underline{2})$$ as our third term, in

middle of the right half:

	1/2	1/1	2/1	
$\underline{0}$ ———————	↓ ———————	↓ ———————	↓	$\underline{1}$
1	2nd	1st	3rd	0

Our first three positive numbers, in this bizarre way of counting, are 1/1, 1/2 and 2/1.

You begin to see why we had to have the goalposts we did: we needed to get 1/1 somewhere along the way, and since numerator and denominator must each be a sum of two other numbers, 0 and 1 - then 1 and 0 - were the only possible choices, if we were to stay positive. And they yielded 1/1 at once.

Now we go on to the third generation, always using our simple rule. Moving from left to right get the new numbers are 1/3 (i.e., $\frac{0+1}{1+2}$), 2/3, 3/2 and 3/1, as you see here:

	1/3	1/2	2/3	1/1	3/2	2/1	3/1	
$\underline{0}$ —	↓ —	↓ —	↓ —	↓ —	↓ —	↓ —	↓	$\underline{1}$
1	4th	2nd	5th	1st	6th	3rd	7th	0

The fourth generation consists of the next eight new middles of each of these eight intervals: from left to right, 1/4, 2/5, 3/5, 3/4, 4/3, 5/3, 5/2, 4/1.

As you continue, always starting at the left and working right, adding adjacent numerators for the new numerator, adjacent denominators for the new denominator, you will get every positive fraction less than 1 in the left half, every positive rational greater than 1 in the right half. Thus 0 and 1 alone, under this remarkable, repeated operation, give rise to every positive rational number. To convince yourself of this, check that 5/3 will be the 13th rational on our list. When will 4/7 come up?

This strange list is called a Farey sequence, after its supposed inventor. But here again - as with le Marquis de l'Hôpital - the history of mathematics hasn't been as precise as mathematics itself. John Farey was an English geologist, who in 1816 published a note on the theorem behind this sequence. He gave it without a proof, perhaps because such was beyond him. It may not, however, have been beyond him to own a copy of a book privately circulated in 1816 by one Henry Goodwyn, who gave both the theorem and its proof. Should we call it henceforth Goodwyn's sequence? No: fourteen years earlier it appeared in France, in an article by a man named Haros, now lost to time. Perhaps Clio, as the muse of history, has taken a sort of ironic retribution in seeing to it that in the <u>Dictionary</u> of <u>National</u> <u>Biography</u> Farey is briefly remembered for his papers (now gathering dust) on the mensuration of timber and on the heights of hills in Derbyshire, and no mention is made of the sequence which alone preserves his name.

Whoever its inventor, this sequence has mind-boggling implications. You'd think that the absence of a smallest positive fraction (you can always make a new one halfway between any candidate and zero) meant that you couldn't count them: there would be no place to start. Put this together with the density of those thickets of fractions which grow between any two integers and it seems only reasonable to conclude that there are many more fractions than counting numbers. Yet the sequence we've just legitimately made shows that they all, mixed in with the whole numbers, can indeed be counted: 1/1 is the first, 1/2 the second, 2/1 the third entry on our list, and so on. Our way of matching them up with the counting numbers may be bizarre, but it achieves its purpose: there are exactly as many positive rational numbers as there are counting numbers, although the latter seem so much thinner on the ground. The uncommon sense of it sends common sense giddily staggering. This is the sort of conduct Cantor uncovered in the Free Republic of Numbers, and the kind of surprise that makes all the struggles of mathematics worth it. The world and the mind are mysterious, but their mysteries are (just barely) accessible.

We have gotten all the rational numbers - and a knock-out insight as well - from 0 and 1. But the question remains, could we get them all from zero solus? We could, were we able to make 0 yield 1, since then we would simply proceed as above. This was the dream of whatever monk it was who wrote the Salem Codex in the twelfth century:

> Every number arises from One, and this in turn
> from the Zero. In this lies a great and sacred
> mystery... He creates all out of nothing, preserves
> and rules it: omnia ex nihilo creat, conservat
> et gubernat.

(Now hold on just a darned minute, as Jimmy Stewart would have said. Adelard of Bath, you remember - also twelfth century - had a student N. O'Creat. But are we dealing instead with some elaborate medieval joke here, a deep pun: the sorcerer's apprentice nihilo creat turning into N. O'Creat? Was the spirit of Nabokov cavorting in Adelard's Bath?)

Tear yourself away from the beckoning vista opened by these parentheses to the yet more alluring prospect of deriving 1 from 0. For with a little ingenuity not God but we humans can do this.

A few painless preliminaries. If you multiply a number n by something and hope to get n as a result, that "something" had better be 1, as you know - the multiplicative identity. Second, a set (call it S) is just a collection of things, and can be broken up into two parts (subsets) - call them A and B - so that whatever was originally in S ends up either in A or in B. Last, the empty set has nothing in it, so it is a subset of any other set (if you have ten marbles in a box, drop a partition in it so that all the marbles are to the right of the divider; then you've made two subsets of this box: the left-hand one is the empty set, the right-hand one has all the marbles). Now we can begin.

Take a heap of Gerbert's apices, on each of which a single number is written, and put them in a box. As you take each one out, multiply the number on it by the numbers on

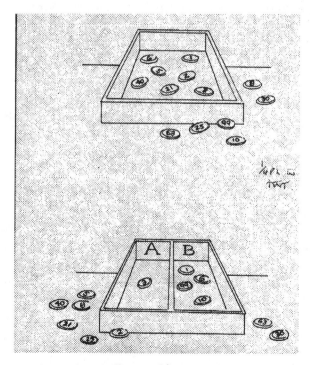

Boxes with counters

those you've already taken out. When you're done, you will have multiplied all these numbers together and gotten some product - call it r. Now put a partition in the box and spill all the counters back in. Some will land in the left compartment (call it A), some in the right (B). Take out the counters from A and multiply the numbers on them together. When you've finished, you will have some product - call it p. Do the same with the counters in B, giving you some product - call it q. Obviously

p·q = r, because you've just done the same multiplication you did before, only broken it into two parts - no matter how many counters were in either compartment. *No matter how many* - even if they had all fallen into B, and A was empty: still p·q = r. But now, since B has all the counters in it, q = r. That means (by our first fact) that p = 1: the product of all the numbers in the set with nothing in it is 1.

Is this too artificial for your taste? Think of it as an example of the recursive abstracting that so deeply stamps mathematical practice: extending the notion of multiplying beyond the realm where it first made sense. This wraps up with one ribbon the glories and despairs of the calling, like taking your understanding out for a vigorous five-mile run - uphill. It was, I've heard, the eminent mathematician John von Neumann who said that in mathematics you don't come to understand things but just get used to them (a thumbnail description of paradigm-shifts and a striking shift itself from d'Alembert's advice just to keep on until faith returns). But while everything recedes to the vanishing point of unprovable axioms, most workers in the field think that proofs should be congenial and what they prove seem grafted onto the stock of your intuition. Mathematics ought to be not only elegant but simple. So let's try a simpler approach to getting everything from nothing, by going back to the idea of the empty set.

Is it an idea or is it out there? I can almost point to it by asking you to consider such things as the set of words beginning with 'xwyz'. As far as the dictionaries and friends I've consulted go, there are no such words: the set is empty. So is the set of all even numbers that are odd, and of all oranges that write heroic couplets. The empty set appears very easy to find if hard to see. But these encounters (if you can call them that, like yesterday upon the stair meeting a man who wasn't there) aren't good enough to guarantee mathematical existence, which can only be deduced from the axioms - in this case, the axioms of set theory - or be one of the axioms themselves. This may seem unnecessarily stringent to you, but if you want mathematics really to be an independent as well as a brave new world, then the creatures in it can't come from anywhere (such as experience) that you would like to have follow from mathematics. And in fact one of the seven axioms of set theory, as it hardened into a reputable branch (some would say the root) of mathematics in the early 1900s, is the axiom that says the empty set exists.

If it bothers you that something - such as the empty set - has to be brought into existence by a mere assertion (and an axiom, for all the confidence with which it is laid down, is no more than that), then notice that the rules of a game or the laws of the land presuppose that there are indeed people to whom they apply. One of the charming idiosyncrasies of mathematics is that these creatures are conjured up along with the principles governing their behavior (and at times it becomes difficult to tell the actor from the action: zero as adjective, noun and verb). They have at their conception no more and no less reality than these principles, but come to seem part of the world as their doings unfold. Looking back from the vantage-point of experiences with them, they stand as foregone premises rather than conclusions.

That said, it must be admitted that the empty set is, as always, an exception. Ernst Zermelo, the first person to think through the axioms of set theory, put his Second Axiom this way in 1908: "There exists a (fictitious) set, the null set, 0, that contains no element at all." Why fictitious? And how do fictitious things 'exist'? I think we're seeing here the same reluctance we met with in Chapter Seven to grant full citizenship to what they call in the North of England 'Offcomeduns' (because they come from 'off'). Perhaps after the seventh generation the family may begin to be thought of as belonging.

If this still leaves you dissatisfied about whether and why the empty set is among the living, perhaps you would be comforted by an argument in the style of St. Anselm's proof of God's existence: this one provided by a philosopher named Wesley Salmon. "The fool saith in his heart," he writes, borrowing Anselm's language, "that there is no empty set. But if that were so, then the set of all such sets would be empty, and hence *it* would be the empty set."

One way or another, then, people have brought the empty set into existence, or found it there. One of the symbols for it is the pair of brackets used for denoting a set, but with nothing between them: { }. Another, Ø, shows how far we've come in a thousand years, since this isn't the medieval phi (ϕ) that sometimes ousted theta (q) for zero, but - if folk etymology is correct - a sign meaning that not even zero is in it: affirming zero's status as a full-fledged number.

Our aim is to show Lear wrong by deriving everything from nothing - that is, from the null set. This trick was pulled off in 1923 by John von Neumann, and it is one we have since gotten used to, although you may feel that for all its simplicity it remains slippery. Identify zero, von Neumann says, with the empty set: it doesn't even have zero in it, it *is* 0. And now (nothing up the sleeves) consider the set that *contains* the empty set: {∅} (or { { } }, if you like that sort of thing). Since this set has one element in it - namely, the empty set - we can identify *it* with the number one. And 2? Why, that is the set containing our previous items: { ∅, {∅} } - and so on, each new set containing all the previous ones, and standing for (or *being)* the next integer.

There! we've gotten all the natural numbers from the empty set, and from these naturals we'll make the negatives, the fractions, the reals and the imaginaries in the well-known ways. "God created the integers, the rest is the work of man", said the 19th century mathematician (and Cantor's enemy) Leopold Kronecker. By von Neumann's argument, the integers too are the work of man, and God - in the spirit of Oken - at most endlessly posited only Himself. "We are the bees of the invisible", Rilke wrote in 1925. Is our task to make the abstract palpable?

You might find von Neumann's prestidigitation faintly disquieting because it has an air to it of angels dancing on the head of a pin. The price we pay for the rigor of modern mathematics is (as we saw in chapters 10 and 11) a certain formality, reminding you that splicing new growths into our intuition isn't the same thing as having them bud from it.

We may just now have walked past the whole point of our story - like a detective who, intent on tracking an obscure clue, brushes against his quarry in the street without knowing until long after. For all our thought, not only the mathematical with its recursive abstracting, is drawn toward formalism, as if our having drained it of the human made it god-like. Only after we've sold our souls to this figure do we realise that it is hollow, adding nothing to what we knew and by multiplying aperçus out to vast generalities, liable to set our understanding at nothing. Is this where the Great Paradigm was leading us - or is formalism rather an occupational hazard of the mind, which is prone to mistake the ever-enlarging context within which content is held for the disappearance of content altogether? So the signs that facilitiate thinking eventually come to be taken for its substance.

Mathematics is a garden whose keeping, however, is in the care of many, tying and trimming their plants that they may naturalise and thrive. The best ligature, in the end, is acknowledging the truth, awkwardly as it may protrude. The "getting used to" von Neumann described is the way understanding rises from stock to stem.

Now "acknowledging truth" comes down to following air-tight deductions. Yet what if a formalist told you that the very logic we use to do this could be wholly derived from denying the truth?

A hair perhaps divides the false and true, in Fitzgerald's translation of Omar Khayyam. What if it doesn't divide but binds them inextricably together? This is in fact what the American philosopher C. S. Pierce showed in a paper of 1880.

The story is curious. Pierce never published his paper, and even if he had it might have been ignored: the lack of recognition he suffered from persisted, he said, because "my damned brain has a kink in it that prevents me from thinking as other people think." What he showed resurfaced with the same symbolism and almost the same form in the work of Henry Maurice Sheffer in 1913, whose monument now is the symbol, the Sheffer Stroke (which we will run into below): a man, you come to think, with a calligrapher's zeal for taking care of the shapes, that the sense might take care of itself. But we'll follow the path that Pierce made.

Let's talk only about declarative sentences, and let's agree that they can have nothing but 0 and 1 as truth-values: they must either be false or true. Then the truth-value of a compound sentence depends on how it is compounded, and on the truth-values of its components. So "Wishes are horses and beggars ride" is true only if both parts are true: if wishes in fact *are* horses and beggars do ride. If either component had truth-value 0, so would the compound. You could if you like show this in a table for the conjunction 'and', with A and B standing for the component sentences and the rows for all possible combinations of ways that A and B could have truth-values 0 and 1:

A	B	A and B
1	1	1
1	0	0
0	1	0
0	0	0

That is the "truth table" for 'and'. You can make the appropriate tables for other ways of conjoining: "A or B", "If A then B", "A if and only if B" and so on ("If wishes were horses then beggars would ride", for example, is only false if the first part is true and the second false: that is, if wishes really are horses yet beggars don't ride).

There are 16 possible arrangements in all of 1s and 0s in our four vertical slots, each corresponding to a different way of compounding two sentences. Pierce showed that each of these could be derived from repeated, judicious use of one: "neither A nor B", which we now symbolise by "A $\downarrow$ B". For example, "if A then B" can be rewritten this way:

$$(((A \downarrow A) \downarrow B) \downarrow ((A \downarrow A) \downarrow B))$$

Since "neither A nor B" is true only when both of its parts are false, it has the truth-table:

A	B	A ↓ B
1	1	0
1	0	0
0	1	0
0	0	1

The whole gamut of compound values begotten by affirming denial! All our assertions reduced to reiterated falsehood! The whole logic of sentences balanced on this negative fulcrum! It is as though we had opened the box in our thinking's innermost sanctum and found 0 in it. An earlier age might well have seen this downward arrow pointing to the abode of The Spirit that Denies, making a hell of heaven and of our demi-paradise.

What Pierce, and Sheffer after him, saw was an elegant technicality: a way of reducing what had previously been the two generators of this logic to one. Does this rank as a deep insight? It was on the order of drawing attention to what was implicitly there, without - so far as I know - remarking on its philosophical implications (but serving to remind us that many shades fill up the spectrum from invention to discovery).

It took a mind with the receptiveness of snow to see the world in such a mere grain of technique. For Wittgenstein this reduction was neither diabolic nor a pretty ornament but the key he had been looking for. With it in hand he recognised that language, built on logic, could only say what *isn't:* but that by sighting along it - looking where it pointed - we could *see* what is; and "Whereof one cannot speak, thereof one must be silent." This was the famous conclusion of his <u>Tractatus Logico-Philosophicus</u>, whose punctiliously numbered sentences stop dead here, followed by an empty expressiveness of paper.

16

THE UNTHINKABLE

Where exactly is it that language ends, beyond which we may make sounds and even shape grammatically correct sentences, but are *saying* nonsense? Consider an earnest pronouncement of the eugenicist Julian Huxley in 1937. He was talking about those born mentally defective, and said that while we should of course give them the best treatment available, it would have been far better for us - and them - had they never been born. For us: you may take exception to his argument but you certainly know what it means (the care, the cost, the anguish we would have been spared). For them? Better for them had they never been born? Which 'them' are we speaking of, and how would they have been better off? Try picturing it, looking on this likeness and on this. Here is Mr. Juke, alive and miserable. And here is Mr. Juke had he never been born, tall, handsome, alert. "Count no man happy until he is dead," said Sophocles. By the obverse of this token, count no man happy until he is born.

The puzzle of non-existence is what has been peeping impishly out at us all through our travels. Now it re-emerges full-grown as we near home. For all our arguments by example, authority or Anselm, how can something as queer as the set of what doesn't exist, itself exist? Thrash about how you will in this nightmare, its images dodge and shift their shapes and return from directions you didn't even know were there, mouthing expressions that, like dividing by zero, are either meaningless or indeterminate. The tautologies and contradictions that flank the operations of logic have no meaning; but how could something within those operations be indeterminate? The answer is that *anything* asserted about what doesn't exist is true. If I say that all octopods have nine tentacles, this is easily shown to be false. If I say they all have eight, that is an empty tautology, since it just repeats in English the meaning of the Greek. But if I tell you that each of the octopods in this room with me has nine tentacles, that is as easily shown to be true, since it is true of every octopod in my room: namely, none. It is also true of my octopods that they *don't* have nine tentacles. The empty set is receding beyond our reach.

If you dismiss logicians as hermetic dreamers and their quibbles as bad dreams, you cannot dismiss an illogical question that comes, at odd moments, to some - and it wouldn't surprise me if it brushed each of us, if only once, with its wing. Looking around, you may wonder at this or that thing in the world: at its origin, when the genealogical fit is on you; at how it works, when the scientific; or why it should be as it is and not otherwise, the good or the ill or the luck of it, when your thought turns philosophic. But what if a moment of Being takes hold and you wonder: why is there anything at all rather than nothing? Why should I, just I, against whose ever existing so many odds were ranged, in fact be alive? And in the Mayan extremity of this mood: why should this frail universe, an ace from nothing, have come to be?

Philosophy begins with wonder, said Aristotle, and it is this particular wonder, according to Schopenhauer, "which keeps the never stopping clock of metaphysics going."

Some have called it not a question but a disease, for which the Germans have the rumbling name "Grübelsucht": moping melancholy mad. Logicians would remind you that it is no more than the tautology: "Why is what is?" - and no less than a tautology, those in its grip remind you.

No rational assault on this ice-wall can succeed, since all attempts to reason about the fittest surviving or chance collisions, the Will or the Would or the Should, occur within its compass. Wittgenstein, who wanted to show the fly its way out of this bottle, gave us his numbered exercises in language to teach us to look not at but along it. But still we stare like a cat at the pointing finger.

So long ago in Greece, when Socrates was young and Parmenides old, the latter laid down a challenge we have sought ever since to pick up. All you can think, he said, is: "Being is." You cannot think non-being, nothing, the void. Using negation, he told us we cannot use negation. All we can think is "Being is." We cannot think motion, change, difference, past or future, here and there, you and me, since each requires thinking 'not'. We can only think: "Being is." How easy to trivialise Parmenides by teaching him to suck eggs: you cannot outlaw negation and proceed to use it. But Parmenides was a poet, and you miss the music if you point out to a poet that his love isn't really a red red rose. Parmenides wanted us to stop talking and listen. Like the background hum from the Big Bang, Being pervades. it fills and *is* the world.

Two millennia later Leibniz heard what he said and recognized with joy the fulness of things. There were no gaps, there was no void, the small became smaller but never nothing. Just as the numbers were choked to bursting with numbers, the world they described was so silted up with beings that it was a continuous whole, a garden whose every leaf was again a garden. Isn't this what the angels sing: "Pleni sunt coeli"? Isn't this the vision rediscovered in our time as fractals - or, in its midnight returns, as a squirming, gasping claustrophobic oppression, the everywhere midden of protista? Both images lie neutrally there for your choosing in the Hindu notion of *brahman,* an omnipresent vitality like salt dissolved in water, like the water on which our sparkling bubbles of self float and burst. The world as empty (sunya), the world as full (asunya): choose which version you like, said Nagarjuna, the master of Mahayana Buddhism, the Great Ferryboat. Opposites are an illusion of language. Something and nothing, you know, are equally false substantives.

Kant too heard Parmenides and felt the strain of Nagarjuna's opposites, and this was his answer. We can't but stitch our impressions together with principles, such as causality, which we take to be out there. The problem is that no matter how fine the weave, it trails off toward ever earlier causes and ever later effects, and our mind wants no sequence of dots to cover these gaps but a unified whole: a picture encircled with a border. So we conjure up a framework we cannot see to satisfy our need for completion. This is the framework of Being, within which our understanding works, nesting our experiences and making sense of them. The completion is illusory - but we would as soon do without it as do without breathing because the air is impure. How can Kant claim this is so? Is he standing in some privileged place outside it? No: his explanation lies within our understanding. It is a mirror made of language, meant to reflect the outside in, just as mathematics, spun out from zero, contains and is contained in its sum.

I write this in the midst of things, in the middle of time. The world extends away on every side, taking its coordinates from a quiet center fitfully seized as self, which, like Wallace Stevens' snowman, listens and beholds

Nothing that is not there and the nothing that is.